THE REEF OPERATING SYSTEM™

Powered by the Equilibrium Method™

Achieving the Holy Grail: A Self-Stabilizing Reef Aquarium

Engineering Natural Balance for Thriving Marine Reef Aquariums

Allan Poulin

Aviator · Systems Thinker · Husband · Father · Reef Steward

Revision A Pre-Publication Proof

Reef Operating System Press

Copyright & Publication Data

The Reef Operating System**TM** Specification: Revision A

Engineering Natural Balance for Thriving Marine Reef Aquariums

Published by Reef Operating System Press.

Paperback ISBN: 979-8-9969713-1-2

Hardcover ISBN: 979-8-9969713-0-5

Premium Digital PDF ISBN: [Optional / To be assigned]

Image & Illustration Notice

Selected visual plates, cover artwork, section-opening images, and aquarium scenes in this Specification include original AI-assisted illustrative imagery created for editorial design purposes.

These images are intended to communicate the beauty, aspiration, structure, and ecological imagination behind the Reef Operating System**TM**. They are not presented as species-identification photography, site-specific reef documentation, or scientific field records.

All diagrams, tables, diagnostic trees, checklists, operating frameworks, and ROS educational materials are original instructional assets created for this Specification.

Where public-domain, licensed, or owner-provided photography is used in future editions, source attribution and usage documentation should be maintained as part of the publication record.

Publication Disclaimer & Responsible Husbandry Notice

The Reef Operating System **TM** Specification is provided for educational, informational, and husbandry-planning purposes.

The methods, frameworks, operating targets, diagrams, checklists, and recommendations presented in this Specification are intended to help aquarists think more clearly about reef aquarium design, biological stability, system operation, and responsible stewardship.

Every aquarium is a unique living system. Outcomes may vary based on aquarium size, livestock selection, water source, equipment, maintenance practices, environmental conditions, experience level, implementation quality, and other factors outside the control of the author or publisher.

No method, including the Reef Operating System **TM**, can guarantee the survival, growth, coloration, reproduction, health, or long-term success of any fish, coral, invertebrate, microorganism, or aquatic ecosystem.

Readers are responsible for researching the needs of all organisms before acquisition; complying with applicable laws, regulations, import restrictions, collection rules, and animal-welfare standards; using electrical equipment, pumps, heaters, lighting, water-treatment systems, chemicals, tools, and aquarium products safely and according to manufacturer instructions; monitoring aquarium conditions appropriately; seeking qualified professional assistance when needed; and acting in the best interest of the living organisms entrusted to their care.

The author and publisher are not responsible for livestock loss, equipment failure, property damage, financial loss, personal injury, or other damages arising from the use, misuse, interpretation, or implementation of the information contained in this Specification.

The Reef Operating System **TM** is a methodology of thoughtful design, observation, measurement, and stewardship. It is not a substitute for judgment, experience, responsible husbandry, or continued learning.

Nature does not maintain coral reefs through constant intervention.
Neither should we.

Dedication

To every aquarist who has ever looked at a thriving coral reef and wondered, "How can I work with nature instead of constantly fighting it?"

This Specification is dedicated to the pursuit of equilibrium.

The Steward's Oath

I will observe before I intervene.
I will measure before I conclude.
I will seek understanding before correction.
I will steward the reef faithfully.

Mission, Vision, and Promise

To make successful reefkeeping more predictable by teaching aquarists how to design ecosystems that naturally maintain the operating characteristics of healthy tropical coral reefs through thoughtful engineering, biological understanding, and responsible stewardship.

Vision: To establish the Reef Operating System**TM** as a universal framework through which reef aquariums are designed, operated, studied, and improved for generations to come.

The ROS Promise: The Reef Operating System**TM** does not promise perfect aquariums. It promises a disciplined framework for making better decisions.

The Constitution of the Reef Operating System**TM**

The Constitution is the philosophical anchor of the Specification. Every Standard should remain consistent with these articles.

ROS Constitution**TM**

Article	Statement
I	Nature is the benchmark.
II	Equilibrium is engineered.
III	Biology comes before technology.
IV	The ecosystem is the primary instrument.
V	Every component must serve a purpose.
VI	Replace what growth consumes.
VII	Stewardship is continuous.
VIII	Every decision should move the ecosystem closer to equilibrium.

About the Author

I have been fascinated by aquariums for as long as I can remember. Like many hobbyists, I began keeping aquariums as a child. What began as curiosity became a lifelong pursuit of understanding how living systems work.

As a pilot, I learned that smooth flights are achieved through disciplined planning, continuous awareness, and small corrections that keep an aircraft on its intended glide path.

My wife, Dr. Jessica Poulin, and I own and manage a maternal-fetal medicine practice together. Through medicine, I have watched individualized care succeed by identifying underlying causes, doing no unnecessary harm, and using the least invasive solution capable of achieving the desired outcome.

Together we are raising five children. Parenting has reinforced the same lessons: consistency creates confidence, routine creates stability, and small daily investments prevent larger problems later.

As a family, we have traveled widely, visited public aquariums across the world, and snorkeled natural reefs in Florida, the Bahamas, Mexico, Africa, and Asia. Everywhere, one observation remained consistent: healthy ecosystems are balanced.

The Reef Operating System**TM** grew from a lifetime of observing systems: aircraft, medical practices, families, businesses, natural reefs, public aquariums, and home aquariums. This Specification is my answer to one question: why do some living systems remain remarkably stable while others require constant correction?

Foreword: A Different Way to Think About Reefkeeping

Every generation of reef aquarists inherits a remarkable gift: decades of observation, experimentation, scientific discovery, and innovation from the generations that came before us.

Never before have aquarists had access to such extraordinary knowledge or such capable equipment. Yet many reef aquariums remain unnecessarily complicated.

The Reef Operating System**TM** was born from a simple question: what if we have been optimizing the wrong thing?

For decades, much of the hobby has focused on controlling the aquarium. ROS asks whether the healthiest aquariums may be those requiring the least unnecessary control because they were designed for equilibrium from the beginning.

This does not mean rejecting technology. Technology should support biology, improve consistency, reduce unnecessary work, and simplify stewardship. It should not become a substitute for understanding how reef ecosystems function.

If you are looking for a list of products to purchase, this is not that book. If you want to understand the principles that allow reef ecosystems to become progressively more stable, predictable, and enjoyable to maintain, welcome to the Reef Operating System**TM**.

Introduction: The Reef Operating System™

An operating system provides the structure that allows many independent components to function together toward a common objective. ROS applies that same philosophy to reef aquariums.

Rather than defining a collection of equipment, ROS establishes a common framework for designing and operating reef aquariums as complete ecosystems.

Every decision is evaluated by one question: does this improve the long-term equilibrium of the ecosystem?

The aquarium is not the project. The ecosystem is. The glass contains it. The equipment supports it. The organisms create it. The aquarist stewards it.

ROS is intentionally manufacturer-neutral. It specifies performance objectives rather than products. This allows the methodology to remain relevant as equipment evolves.

The Specification is scale independent. A desktop reef, family room display, office aquarium, coral farm, and public aquarium may implement ROS differently while remaining faithful to the same operating principles.

Each Standard includes purpose, philosophy, figures, ROS Insights™, Field Notes from Allan™, common mistakes, verification, a decision framework, and key takeaways.

Table of Contents

Nature does not maintain coral reefs through constant intervention. Neither should we.

STANDARD 1

Foundations of Ecosystem Design

Design for Equilibrium. Operate Through Stewardship. Improve Through Observation.

Purpose

The purpose of this Standard is to establish the philosophical and engineering foundation of the Reef Operating System**TM** (ROS). Every decision made throughout this Specification flows from the principles introduced here.

ROS is not defined by a specific filter, lighting fixture, dosing product, or brand of equipment. It is defined by a way of thinking.

The central premise of ROS is that a reef aquarium should be designed so biological, chemical, and physical processes naturally support one another.

Why This Standard Matters

Modern reefkeeping offers more equipment, information, and products than ever before. These tools are valuable, but they can also encourage aquarists to treat symptoms rather than design for stability.

ROS proposes a different approach: engineer equilibrium instead of managing instability.

The First Principle

Engineer equilibrium instead of managing instability. The aquarium platform, Living Substrate**TM**, lighting, water movement, gas exchange, livestock, nutrition, automation, and monitoring exist to help the ecosystem naturally maintain the operating characteristics of a healthy tropical reef.

The Three Pillars of ROS

Design for Equilibrium**TM**: build stability from the first decision.

Operate Through Stewardship**TM**: observe, verify, and intervene intentionally.

Improve Through Observation**TM**: let the reef teach the aquarist.

Figure 1-1: The Three Pillars of ROS

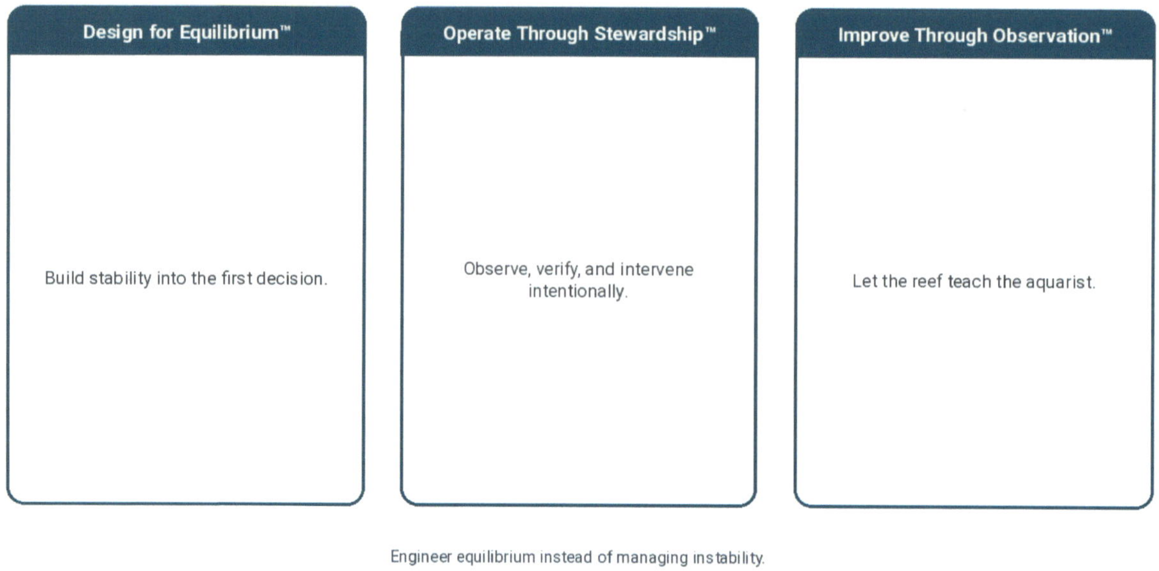

Figure 1-1: The Three Pillars of ROS.

ROS Insight™
Systems are greater than their parts. Equilibrium emerges when ordinary processes support one another in balance. That is why ROS is an operating system rather than a filtration method.

Field Note from Allan™
Through flying, medicine, parenting, business, travel, and a lifetime in the aquarium hobby, I saw the same lesson repeated: healthy systems rarely require dramatic correction. They are designed well, observed carefully, and maintained through small, deliberate adjustments.

Common Mistakes

- Beginning equipment purchases before defining biological goals.
- Treating water chemistry as independent from biology.
- Assuming more equipment automatically produces greater stability.
- Making multiple adjustments simultaneously.
- Chasing parameters instead of evaluating the ecosystem.

Verification

- Every equipment choice supports a defined biological objective.
- The aquarium is designed as an integrated ecosystem.
- Simplicity is preferred until evidence supports added complexity.
- Decisions consistently move the ecosystem toward Operating Targets**TM**.

Decision Framework TM

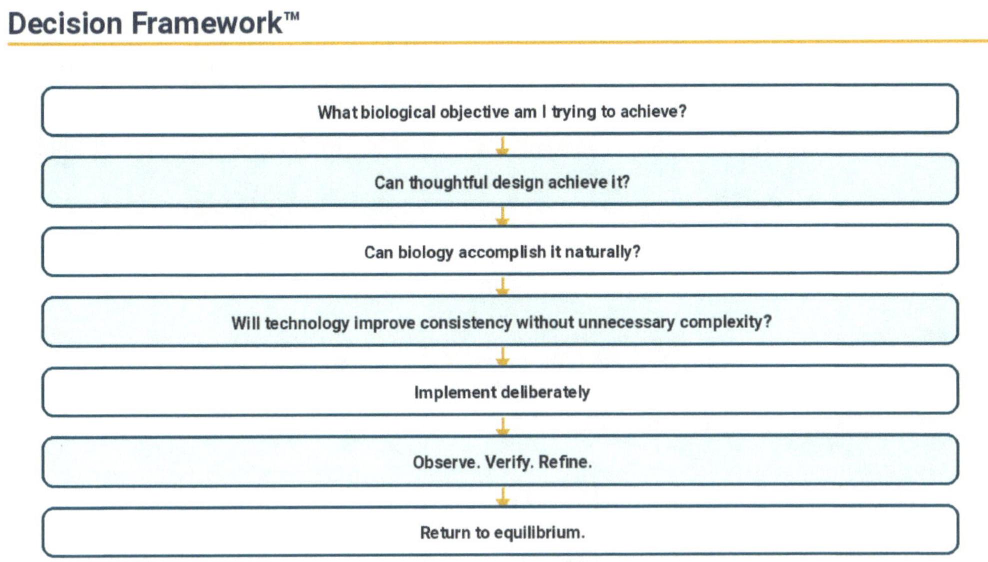

Decision Framework™

Key Takeaways

- ROS is an operating philosophy before it is a collection of techniques.
- Healthy reef aquariums are engineered as integrated ecosystems.
- Equilibrium is designed in from the beginning.
- Stewardship replaces continual intervention.
- Observation and measurement guide improvement.

ROS Principle™
Design for Equilibrium. Operate Through Stewardship. Improve Through Observation.™

Standard 2: Aquarium Platform Selection

The best aquarium is not the most expensive one. It is the one that best supports the ecosystem you intend to build.

STANDARD 2

Aquarium Platform Selection

Selecting the Foundation of the Ecosystem

Purpose

The purpose of this Standard is to establish the engineering principles used to select the physical platform upon which ROS is built.

The aquarium is the structural foundation of a living ecosystem. Every decision after this point is influenced by the platform selected.

Why This Standard Matters

Many aquarists begin by asking which aquarium to buy. ROS begins by asking what ecosystem the aquarium must support.

The goal is not the largest or most expensive platform. The goal is the most appropriate platform.

The ROS Platform Philosophy

ROS avoids dependence upon proprietary systems. Commercially available aquariums provide broadly compatible platforms, widely available accessories, reliable stands, lids, lighting options, and replacement parts.

Lids as Passive Stability

A properly fitted lid reduces evaporation, improves temperature stability, reduces salt creep, lowers top-off demand, limits humidity, and helps prevent fish jumping.

Simplicity Creates Opportunity

Sumps, skimmers, controllers, and external plumbing are compatible with ROS when they solve a measurable problem. They are not prerequisites.

Figure 2-1: Aquarium Platform Selection Matrix

Criterion	Engineering Objective
Volume	Supports intended livestock
Access	Maintenance without frustration
Compatibility	Standard equipment support
Stability	Lid, ATO, temperature control
Future Growth	Room for maturity

Figure 2-1: Aquarium Platform Selection Matrix.

Platform Selection Checklist

Criterion	ROS Question	Verification
Biological Fit	Does the platform support the intended livestock?	Adult size, swimming space, coral growth room
Access	Can it be maintained comfortably for years?	Reach, viewing height, stand ergonomics
Compatibility	Can standard equipment support it?	Lighting, lids, ATO, heaters, pumps
Stability	Does it support passive stability?	Lid, temperature control, top-off
Future Growth	Will it remain suitable as the reef matures?	Room for colonies and maintenance

ROS Insight™

Infrastructure should serve biology. A simple, commercially available aquarium designed around biological objectives can outperform a more expensive system designed around complexity.

Field Note from Allan™

After visiting public aquariums around the world, I noticed that the healthiest displays make the equipment disappear. The livestock behaves naturally and the ecosystem feels balanced. ROS scales that lesson from public aquariums to home and office reefs.

Common Mistakes

- Purchasing the largest aquarium without considering maintenance.
- Designing around equipment instead of biology.
- Underestimating access.
- Ignoring the value of lids and passive stability.
- Assuming cost alone determines performance.

Verification

- The platform supports intended livestock and ecosystem goals.
- Maintenance is practical.
- Commercial equipment is readily available.
- The platform allows future growth and modification.

Decision FrameworkTM

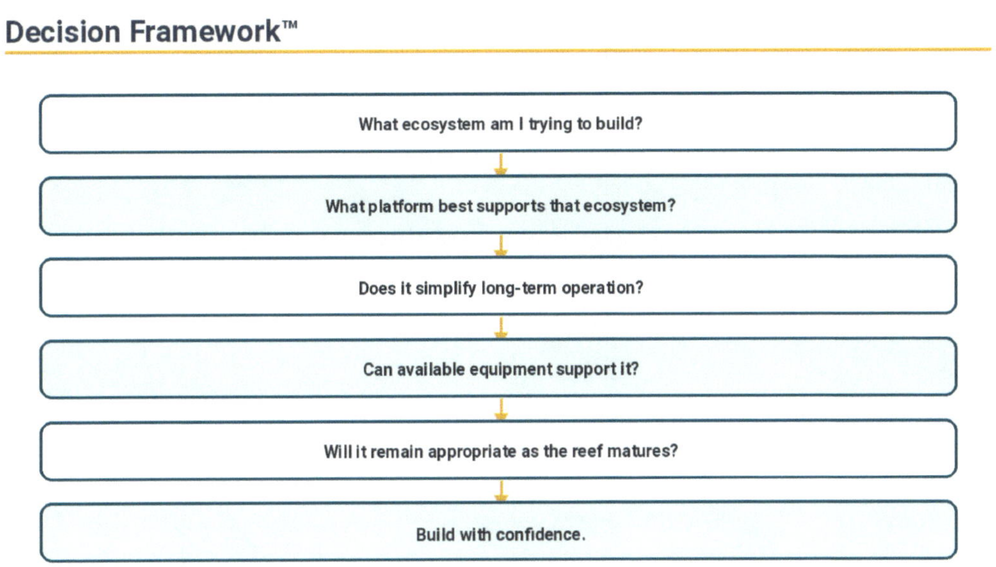

Decision Framework™

Key Takeaways

- Select the aquarium to support the ecosystem.
- Commercial aquariums often provide outstanding ROS platforms.
- Simplicity increases flexibility.
- Infrastructure should support biology.
- ROS scales from nano aquariums to large custom installations.

ROS Principle™
Invest in Biology, Not Infrastructure.™

The ocean does not measure success by how much rock it contains. It measures success by how well life uses the space around it.

STANDARD 3

Ecosystem ArchitectureTM

Designing Space for Biology, Not Just Rock

Purpose

The purpose of this Standard is to establish the architectural principles governing the physical layout of a Reef Operating SystemTM.

Aquascaping is environmental engineering. Rock, substrate, open water, and coral placement influence circulation, gas exchange, nutrient cycling, fish behavior, and long-term stability.

Why This Standard Matters

A common tendency is to fill the aquarium with more rock, coral, fish, and equipment. ROS asks how much space the ecosystem needs to function naturally.

Functional Open WaterTM

Open water is not empty space. It provides swimming room, broad circulation, nutrient transport, visual depth, room for coral growth, and maintenance access.

The Patch Reef Philosophy

ROS generally favors distinct reef structures separated by open water and visible substrate. Build an ecosystem that breathes.

Design for the Future

Plan for colonies, not frags. Each coral should be placed with mature size, growth form, and competitive behavior in mind.

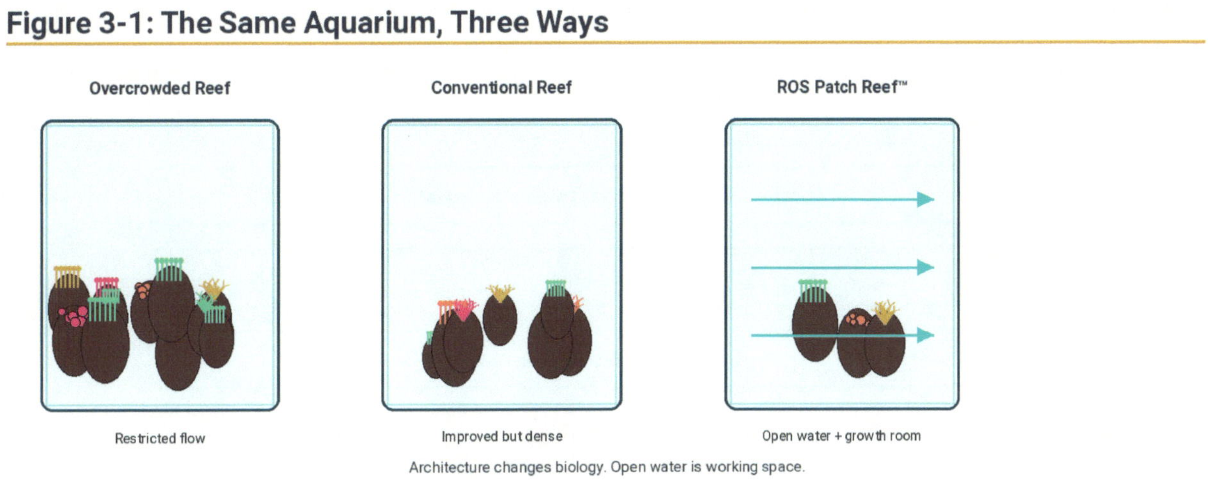

Figure 3-1: The Same Aquarium, Three Ways.

ROS Insight™

Empty space is working space. Within ROS, Functional Open Water™ is one of the most productive areas of the aquarium because it allows life, light, and water to move naturally.

Field Note from Allan™

While snorkeling and visiting aquariums, my family and I noticed that healthy reefs rarely looked crowded. Water moved freely, fish had room, and coral colonies could grow into their natural forms. A young reef should have room to become an old reef.

Common Mistakes

- Filling the aquarium with unnecessary rock.
- Designing for frags instead of colonies.
- Sacrificing swimming space for visual density.
- Creating inaccessible detritus traps.
- Treating open water as wasted space.

Verification

- Water circulates freely around major structures.
- Fish have unobstructed swimming areas.
- Corals have room to mature.
- Maintenance can be performed without dismantling the aquascape.

Decision Framework™

Decision Framework™

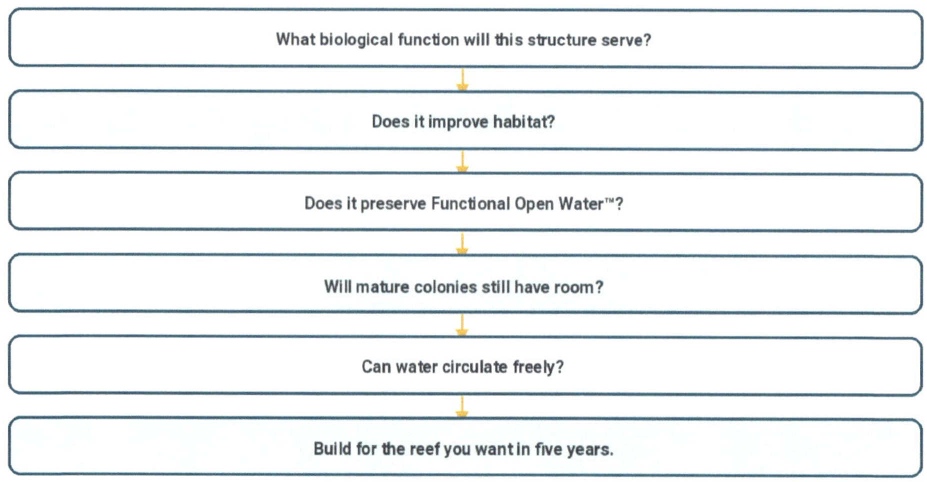

Decision Framework™

Key Takeaways

- Aquascaping is ecosystem engineering.
- Functional Open Water™ is essential, not unused.
- Patch reef architecture promotes circulation and growth.
- Rock should support biology rather than dominate the aquarium.
- The reef should become more beautiful with time.

ROS Principle™
Build the framework. Let biology create the masterpiece.™

The reef is built from the bottom up.

STANDARD 4

The Living SubstrateTM

Engineering the Biological Foundation of the Reef

Purpose

The purpose of this Standard is to establish the biological and engineering principles governing the substrate within ROS.

The substrate is one of the most important biological components of the ecosystem. It provides microbial habitat, nutrient transformation, chemical compatibility, and a stable interface between biology and engineering.

Why This Standard Matters

Many aquariums treat substrate as optional or aesthetic. ROS treats it as living infrastructure that should be designed, commissioned, and stewarded.

The Living SubstrateTM

The Living SubstrateTM combines aragonite substrate, uniform under-substrate circulation, microbial communities, benthic life, and nutrient exchange.

Engineering the Biological Foundation

The preferred implementation uses a modified undergravel filter designed to provide roughly 50 gallons per hour of distributed flow across the bottom. Uniformity of flow is more important than the exact number.

Passive Stability Through Aragonite

Aragonite is a passive safety net. ROS does not rely on aragonite alone to maintain pH or alkalinity, but it contributes chemical compatibility within a broader system.

Substrate Depth

ROS generally favors 1/2 to 1 1/2 inches of aragonite substrate. Deeper beds may be appropriate when justified by livestock or aesthetics.

Figure 4-1: The Living Substrate™ Cross-Section

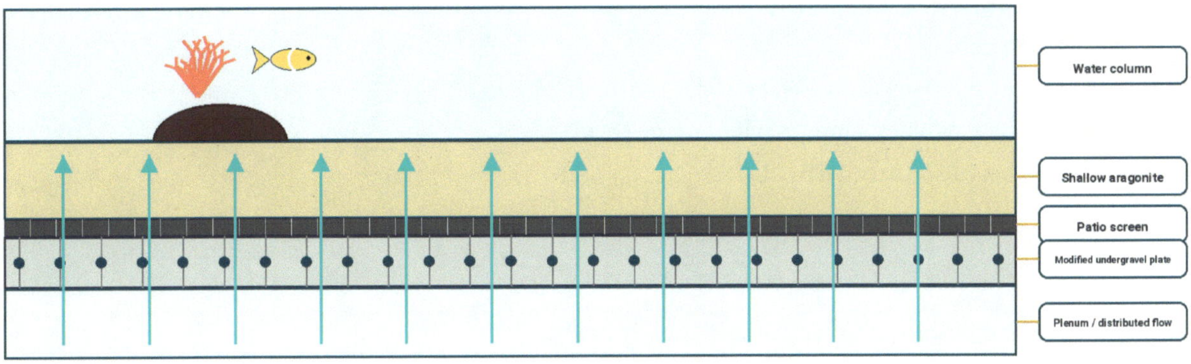

The substrate is engineered living infrastructure, not decoration.

Figure 4-1: The Living Substrate™ Cross-Section.

Substrate Design Reference

Component	Preferred ROS Guidance	Reason
Material	Aragonite-based media	Chemical compatibility with reef systems
Depth	1/2 to 1 1/2 inches	Surface area with manageable stewardship
Barrier	Patio screen for fine media	Protects UGF plate from fine substrate
Cleanup Crew	Match species to grain size	Function-based selection
Flow	Uniform distributed flow	Supports predictable biology

ROS Insight™

Biology lives at interfaces. Water meeting substrate is one of the most important interfaces in the reef aquarium. ROS engineers this interface intentionally.

Field Note from Allan™

Modern reefkeeping largely abandoned one of the largest biological surfaces in the aquarium: the bottom. Once I began seeing the substrate as living infrastructure rather than decoration, many pieces of ROS fell into place.

Common Mistakes

- Selecting substrate based solely on appearance.
- Using excessive depth without objective.
- Allowing stagnant zones.
- Randomly selecting clean-up crew species.
- Assuming maturity ends when ammonia and nitrite reach zero.
- Disturbing the substrate unnecessarily.

Verification

- The substrate remains biologically active and visually healthy.
- Uniform flow exists beneath the substrate.
- The clean-up crew matches the substrate.
- Nutrient processing becomes predictable.

Decision Framework TM

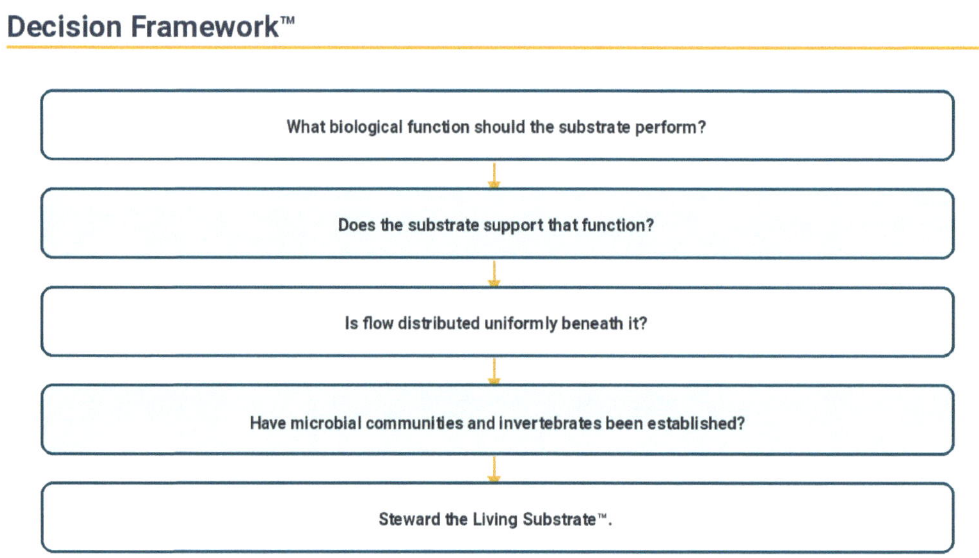

Decision Framework™

Key Takeaways

- The substrate is living infrastructure.
- The Living Substrate TM is a primary biological engine.
- Uniform under-substrate flow supports long-term function.
- Aragonite contributes passive compatibility.
- Biological succession continues beyond cycling.

ROS Principle™

The strongest reefs are built from the bottom up.™

Corals evolved beneath the sun—not beneath a brand.

STANDARD 5

Lighting & Energy Delivery™

Reproducing the Sun, Not Selling the Fixture

Purpose

The purpose of this Standard is to establish the principles governing light as the primary energy source within ROS.

Lighting is selected because it delivers the quantity, quality, distribution, and duration of light required by the intended ecosystem.

Why This Standard Matters

Every ecosystem is driven by energy. In tropical reefs, light influences photosynthesis, calcification, pigmentation, growth, competition, nutrient demand, microbial activity, and animal behavior.

The ROS Lighting Philosophy

The fixture is simply a delivery system. Performance, not price or prestige, is the objective.

Mimic Nature

ROS favors gradual sunrise, stable daylight, gradual sunset, and predictable daily timing.

PAR Before Opinion

Lighting should be verified with a PAR meter at the upper reef, mid-water, sand bed, and intended coral zones.

Energy Drives Demand

Increasing light increases biological demand. Lighting changes should be deliberate and observed.

Figure 5-1: The ROS Energy Map™

Different energy zones allow different organisms to occupy the conditions suited to their biology.

Figure 5-1: The ROS Energy Map™.

PAR Mapping Reference

Zone	Typical Use	ROS Guidance
Upper Reef	High-light corals	Measure and acclimate gradually
Mid-Water	Mixed coral placement	Use as primary design zone
Sand Bed	Lower-light corals / open substrate	Verify before placing livestock
Shaded Areas	Habitat diversity	Useful when intentional

ROS Insight™

The coral responds to photons, not marketing. The objective is an appropriate energy landscape for the livestock being maintained.

Field Note from Allan™

One liberating realization was that the aquarium does not know what brand of light hangs above it. It experiences the light itself. That shifted my focus from brands to measured performance.

Common Mistakes

- Selecting lighting based on popularity.
- Assuming identical settings work for every aquarium.
- Changing intensity frequently.
- Ignoring PAR distribution.
- Designing livestock around the light instead of the reef.

- Equating higher intensity with better performance.

Verification

- PAR has been measured and documented.
- Lighting matches intended livestock.
- Schedules remain consistent.
- Coral color and growth appear natural.
- Biological performance remains stable.

Decision Framework TM

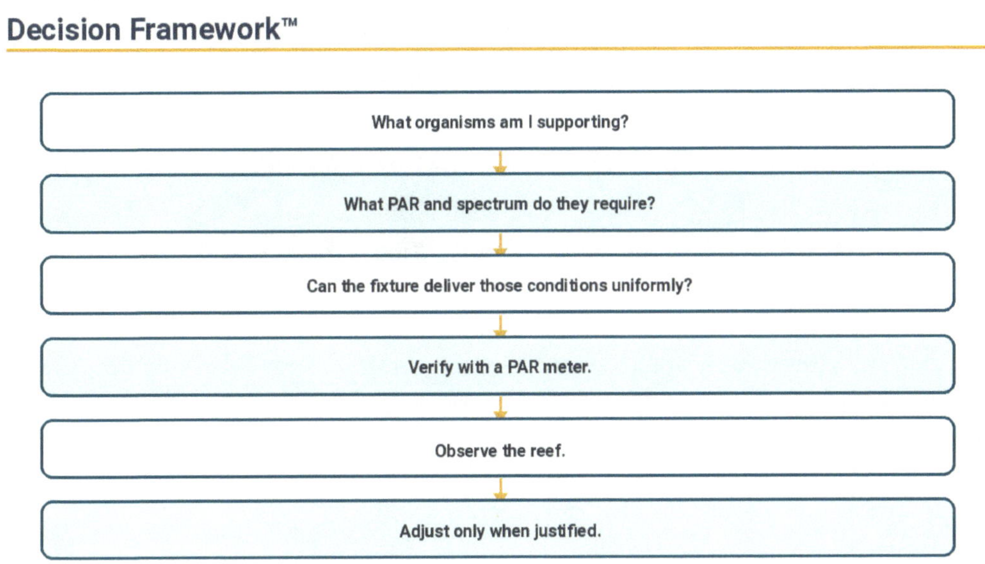

Decision Framework™

Key Takeaways

- Lighting is the primary energy source.
- The objective is to reproduce tropical sunlight functionally.
- PAR should be measured.
- Stable lighting beats constant changes.
- Increased light increases biological demand.
- Performance matters more than price.

ROS Principle™
Measure the light. Observe the reef. Let biology determine success.™

Water movement is not about creating current. It is about delivering life.

STANDARD 6

Water Movement & Gas Exchange™

Engineering Circulation to Support Biology

Purpose

The purpose of this Standard is to establish the principles governing water movement and gas exchange within ROS.

Circulation transports oxygen, carbon dioxide, nutrients, dissolved elements, food, waste, and heat. Gas exchange supports the operating characteristics of a healthy tropical reef.

Why This Standard Matters

Every organism depends on moving water. Flow connects the ecosystem, while gas exchange links the aquarium to the atmosphere.

The ROS Flow Philosophy

Flow should deliver oxygen, transport nutrients, remove waste, support gas exchange, reach functional areas, and support livestock.

Gas Exchange

Oxygen enters and carbon dioxide leaves. Effective gas exchange influences respiration, pH, calcification, and biological performance.

The Role of the CO_2 Scrubber

A CO_2 scrubber is a preferred tool when elevated indoor CO_2 limits pH. It can supply conditioned air through an air stone, air pump, skimmer, or air-driven UGF.

Protein Skimmers

Skimmers are compatible tools, not universal requirements. The question is whether the aquarium benefits based on biological demand and operating objectives.

Figure 6-1: The Air-Water-Biology Cycle™

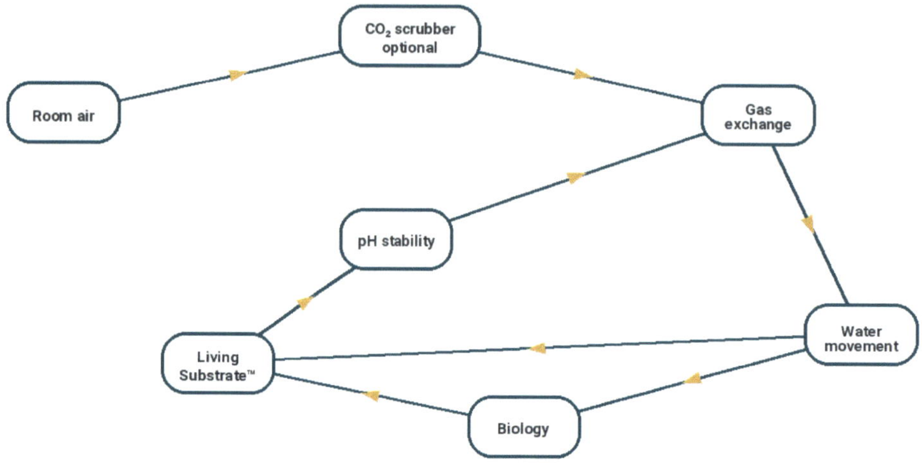

Gas exchange, pH, respiration, and circulation are one integrated operating system.

Figure 6-1: The Air-Water-Biology Cycle™.

Flow & Gas Exchange Reference

Objective	Preferred ROS Approach	Verification
Oxygen delivery	Broad circulation + air exchange	Fish respiration, coral extension
CO2 management	Improve air quality, consider scrubber	Stable pH trend
Substrate support	Distributed UGF flow	Active Living Substrate™
Coral support	Flow matched to organisms	Natural growth and polyp extension

ROS Insight™

Water carries more than water. It carries oxygen, carbon dioxide, food, waste, heat, minerals, larvae, and microorganisms. Flow is biology in motion.

Field Note from Allan™

I stopped asking how many gallons per hour my pumps produced and began asking whether every part of the ecosystem was receiving what it needed. Pump placement became more important than pump size.

Common Mistakes

- Selecting pumps based on advertised flow alone.
- Creating excessive flow in some areas and stagnation in others.

- Assuming more flow is always better.
- Ignoring gas exchange.
- Using equipment without a biological objective.
- Treating CO_2 only as a chemistry problem.

Verification

- Water reaches all functional areas.
- Surface gas exchange is effective.
- Corals show healthy extension.
- No persistent stagnant zones are observed.
- The Living Substrate**TM** receives uniform distributed flow.

Decision Framework**TM**

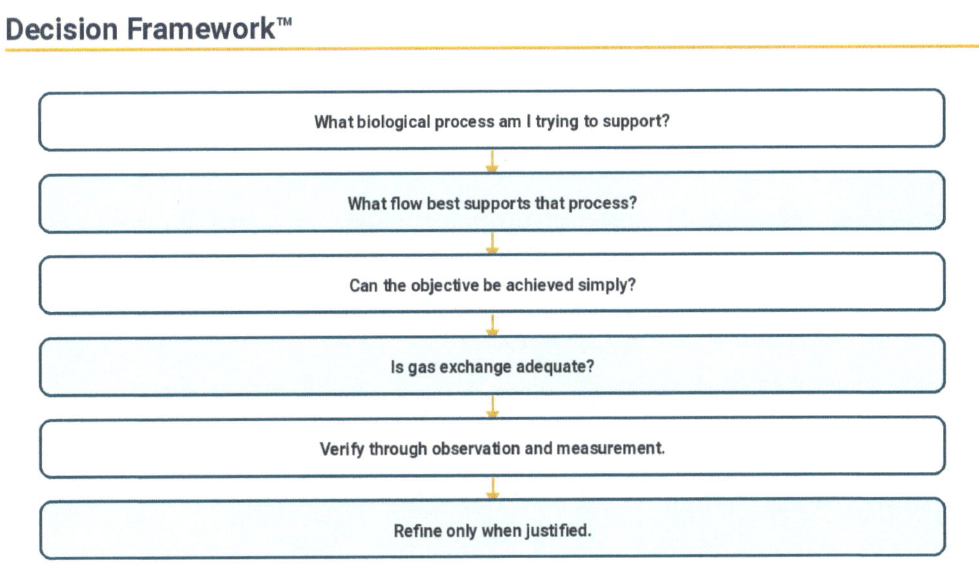

Decision Framework™

Key Takeaways

- Water movement exists to support biology.
- Gas exchange is fundamental to equilibrium.
- Flow should be designed around organisms.
- CO_2 management begins with incoming air.
- Skimmers are tools, not requirements.
- Simplicity is preferred when effective.

ROS Principle™

On a natural reef, fish are not separate from the corals. They help feed them.

STANDARD 7

Feeding the Ecosystem™

Nutrition as the Engine of Biological Equilibrium

Purpose

The purpose of this Standard is to establish the nutritional philosophy of ROS.

Feeding is the primary method by which energy, nutrients, and essential elements enter the ecosystem. Every feeding event influences fish, corals, microbes, nutrient cycling, and long-term balance.

Why This Standard Matters

Food is not simply nutrition for individual animals. It is energy entering the ecosystem. Fish consume food, microbes process waste, and corals benefit from nutrient cycling.

The ROS Feeding Philosophy

ROS generally prefers multiple smaller feedings over fewer large feedings when practical. Consistency supports equilibrium.

Food Quality Matters

Modern prepared foods can form an excellent foundation. Variety may include pellets, frozen foods, fresh foods, species-specific foods, and targeted feeding where justified.

Fish Feed the Reef

Fish contribute directly to reef health by metabolizing food and participating in nutrient cycling. This supports the broader nutrient economy when balanced responsibly.

Automation Improves Consistency

Automatic feeders can deliver frequent small meals with precision. Automation improves repeatability; it does not replace observation.

Figure 7-1: Nutrient Flow in ROS

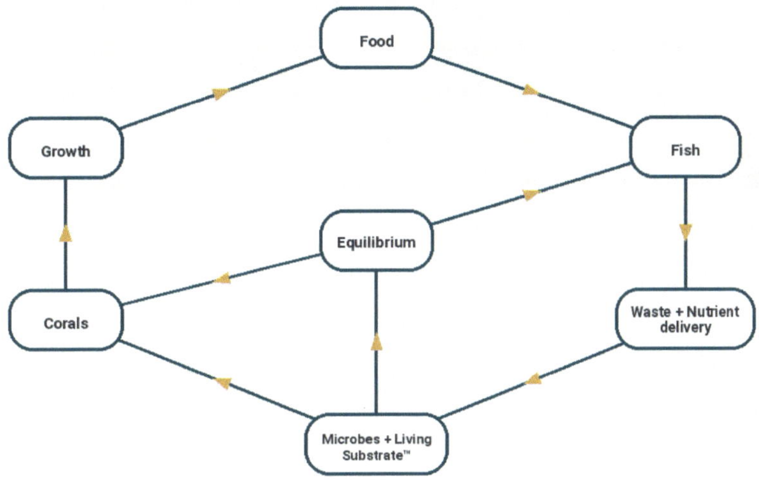

Food is not waste. Food is energy entering the ecosystem.

Figure 7-1: Nutrient Flow in ROS.

Feeding Portfolio Example

Food Type	Role	Operational Note
Pellet foods	Consistent automated foundation	Blend complementary formulas
Frozen foods	Dietary diversity	Use intentionally, not randomly
Fresh marine foods	Occasional enrichment	Appropriate handling required
Target foods	Specialized organisms	Use when biologically justified

ROS Insight™

Nutrients are resources to manage responsibly, not enemies to eliminate. The objective is neither starvation nor excess. The objective is balance.

Field Note from Allan™

I began thinking about nutrient delivery over the course of an entire day rather than breakfast and dinner. Automatic feeders helped create a steady nutritional glide path for the reef.

Common Mistakes

- Feeding based on convenience rather than biology.
- Large infrequent feedings that create nutrient swings.

- Using one food exclusively.
- Overfeeding to accelerate growth.
- Underfeeding out of fear.
- Adjusting feeding without observing the ecosystem.

Verification

- Fish maintain excellent body condition.
- Feeding behavior is vigorous and natural.
- Corals show healthy coloration and growth.
- Nutrient trends remain stable.
- Feeding schedules are predictable.

Decision Framework™

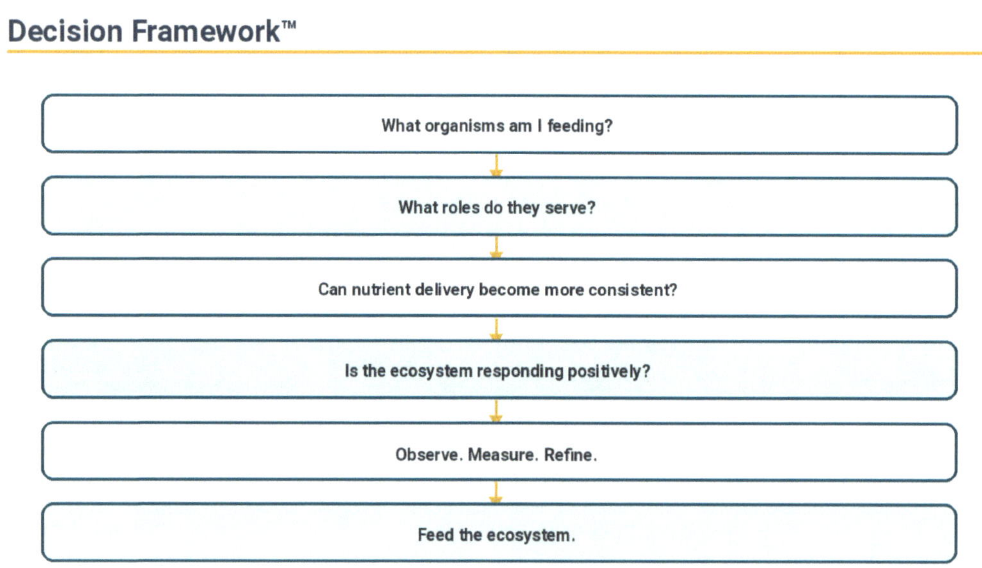

Decision Framework™

Key Takeaways

- Feeding introduces energy and nutrients.
- Fish participate in nutrient cycling.
- Smaller frequent feedings align with natural ecology.
- Automation improves consistency.
- Nutrition evolves as the reef matures.
- Observation measures success.

ROS Principle™

Standard 8: Operations & Stewardship

The greatest improvements to a mature reef often come from knowing when not to interfere.

STANDARD 8

Operations & Stewardship™

Operating the Reef Rather Than Managing Problems

Purpose

The purpose of this Standard is to establish the day-to-day operational philosophy of ROS.

Once the aquarium is commissioned and equilibrium begins to develop, the aquarist's role changes from construction to stewardship.

Why This Standard Matters

Many reefs become unstable because they experience continual change. ROS encourages consistent operation so the ecosystem spends more time functioning than recovering.

The Stewardship Mindset

Stewardship requires observation, consistency, patience, documentation, and deliberate decision-making.

Operational Monitoring™

Monitoring verifies that the ecosystem remains on its Glide Path™ toward Operating Targets™. It combines observation with targeted measurements.

Build Predictable Routines

Consistent feeding, supplementation, inspection, cleaning, monitoring, and ICP verification create stability.

Small Corrections Preserve Stability

When changes are necessary, adjust one variable, observe the response, and document the outcome.

Figure 8-1: The ROS Dashboard™

A preflight checklist for stewardship: brief, routine, and disciplined.

Figure 8-1: The ROS Dashboard™.

Operational Monitoring Schedule

Frequency	ROS Task	Purpose
Daily	Visual observation, feeding, equipment glance	Stewardship awareness
Weekly	Test strip or routine monitoring, salinity check	Confirm glide path
Monthly	Alkalinity verification, equipment cleaning	Trend awareness
Quarterly	ICP, review growth, inspect flow and PAR if needed	Long-term verification
Annual	Full operating system review	Continuous improvement

ROS Insight™

Quiet reefs are healthy reefs. Predictability is one of the strongest indicators of equilibrium.

Field Note from Allan™

Flying taught me that smooth flights require preparation, monitoring, and small corrections. When reef systems are well designed, daily operation becomes calm. I ask whether the reef is continuing to tell me it is healthy.

Common Mistakes

- Making changes too frequently.
- Adjusting multiple variables simultaneously.
- Reacting to isolated results.
- Failing to document observations.
- Mistaking activity for progress.
- Believing every deviation requires immediate correction.

Verification

- Maintenance follows a consistent routine.
- The ecosystem remains predictable.
- Interventions become less frequent.
- Records demonstrate stability.
- The aquarist spends more time observing than correcting.

Decision FrameworkTM

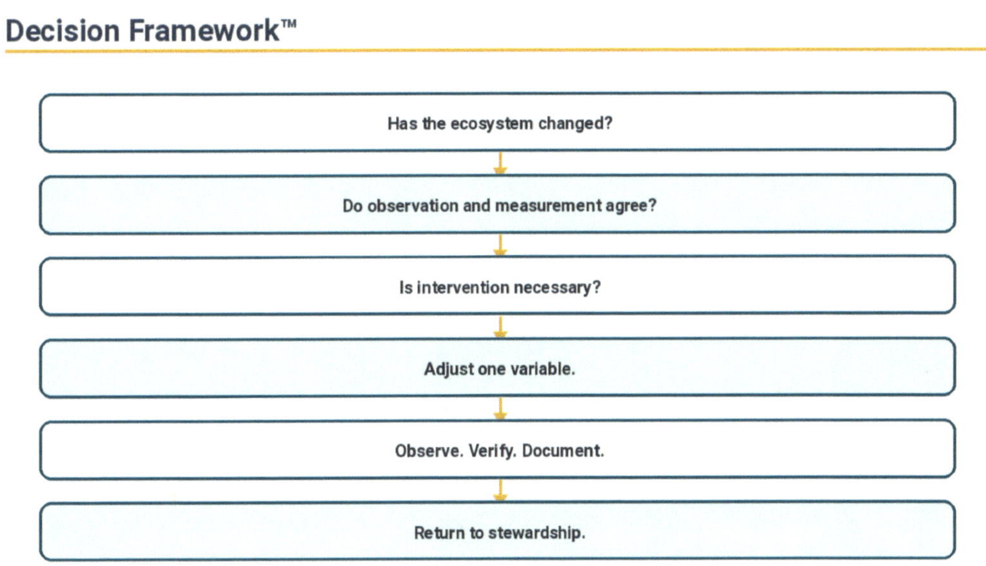

Decision Framework™

Key Takeaways

- A commissioned reef should spend most of its life operating.
- Stewardship emphasizes consistency and restraint.
- Operational MonitoringTM confirms the Glide PathTM.
- Documentation creates knowledge.
- Small corrections preserve stability.

- The goal is predictability and resilience.

ROS Principle™

Operate with consistency. Intervene with intention. Steward with patience.™

A successful reef is not defined by the rarity of its inhabitants, but by how well they thrive together.

STANDARD 9

Livestock Selection & Ecosystem Compatibility™

Building Communities Rather Than Collections

Purpose

The purpose of this Standard is to establish the principles governing livestock selection within ROS.

Livestock is not selected only because it is attractive. Each organism becomes a participant in the ecosystem.

Why This Standard Matters

Incompatible livestock can destabilize even well-equipped aquariums. ROS asks whether an animal can be provided with appropriate habitat throughout its life.

The Reef Comes First

The ecosystem determines livestock. Additions should be gradual so the reef can adapt to changing nutrient demand, competition, feeding requirements, and behavior.

Functional Roles™

ROS organizes livestock according to ecological function: Nutrient Processors™, Algae Managers™, Sand Surface Maintainers™, Planktivores™, Architectural Corals™, and Habitat Specialists™.

Design for Adult Animals

Evaluate adult size, behavior, feeding requirements, environmental needs, and long-term compatibility before purchase.

Quarantine & Biosecurity

Every aquarist should have a deliberate protocol for minimizing disease, parasite, and pest risk.

Figure 9-1: Livestock by Functional Role

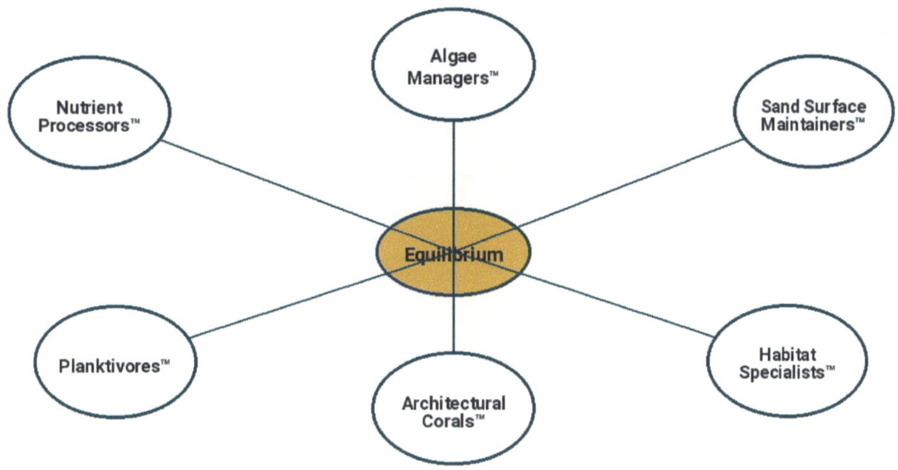

Select organisms by ecological function, not impulse.

Figure 9-1: Livestock by Functional Role™.

Functional Livestock Roles

Role	Examples	ROS Function
Nutrient Processors™	Fish with regular feeding/grazing behavior	Deliver nutrients through the ecosystem
Algae Managers™	Tangs, blennies, snails, urchins	Graze and control algae pressure
Sand Surface Maintainers™	Cerith snails, selected substrate animals	Maintain Living Substrate™ surface
Planktivores™	Anthias, chromis, selected wrasses	Support regular nutrient input
Architectural Corals™	Branching, plating, encrusting colonies	Define mature reef structure
Habitat Specialists™	Species with precise needs	Only when ecosystem can support them

ROS Insight™

Every animal changes the ecosystem. Thoughtful selection protects equilibrium; impulsive selection often disrupts it.

Field Note from Allan™

Traveling and observing animals in natural habitats changed the way I stock aquariums. I rarely ask only whether a

fish or coral is beautiful. I ask whether it belongs in the ecosystem I am building.

Common Mistakes

- Purchasing livestock impulsively.
- Designing around juvenile size only.
- Overcrowding.
- Ignoring ecological roles.
- Adding multiple animals at once.
- Underestimating biosecurity.

Verification

- Livestock exhibit natural behavior.
- Growth remains appropriate.
- Feeding competition is minimal.
- The community remains compatible.
- New additions are deliberate.

Decision FrameworkTM

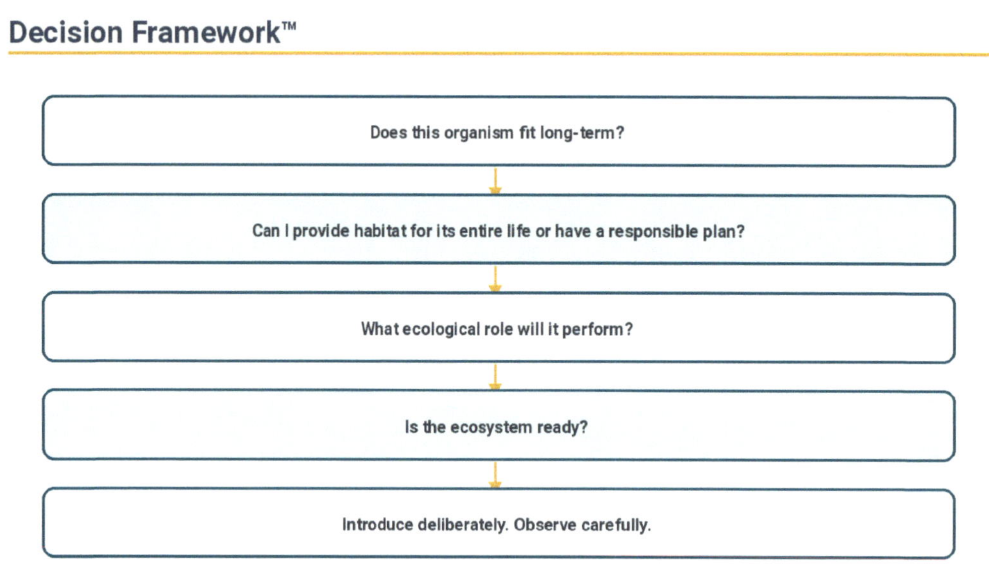

Decision Framework™

Key Takeaways

- Build communities, not collections.
- Select livestock by function as well as beauty.

- Plan for adult size and lifelong care.
- Introduce organisms gradually.
- Protect the ecosystem through biosecurity.
- Responsible restraint is stewardship.

ROS Principle™
Choose organisms that strengthen the ecosystem—not simply the display.™

The nitrogen cycle starts the aquarium. Biological succession builds the reef.

STANDARD 10

Ecosystem Commissioning™

From Sterile Aquarium to Living Reef

Purpose

The purpose of this Standard is to establish the procedures and philosophy governing the commissioning of ROS.

Commissioning is the deliberate process of transforming a new aquarium into a functioning ecosystem. It is more than completing the nitrogen cycle.

Why This Standard Matters

Cycling is a milestone. Commissioning is the complete process of establishing biological, chemical, physical, and operational readiness.

Phase 1 — System Construction

Assemble platform, Living Substrate™, under-substrate circulation, aquascape, lighting, flow, temperature control, gas exchange, and initial chemistry.

Phase 2 — Fishless Biological Cycling

Use an appropriate ammonia source. Keep lighting off or minimal. Monitor chemistry closely while the Living Substrate™ begins colonization.

Phase 3 — Biological Seeding

Encourage diversity through microbial cultures, live rock, copepods, amphipods, beneficial invertebrates, and appropriate inoculation methods.

Phase 4 — Initial Livestock

Introduce hardy livestock gradually under deliberate biosecurity protocols. Let the ecosystem determine readiness.

Phase 5 — Nutrient Base and Light

Establish feeding, nutrient processing, and lighting gradually. Increasing light increases biological demand.

Figure 10-1: ROS Commissioning Timeline™

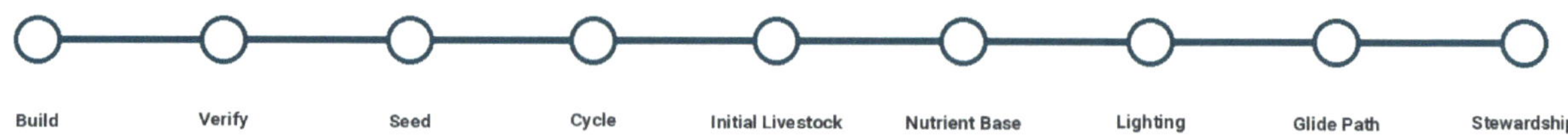

Biological milestones determine progress. The calendar does not commission a reef.

Advance only when the ecosystem demonstrates readiness.

Figure 10-1: The ROS Commissioning Timeline™.

Commissioning Phase Checklist

Phase	Primary Objective	Advance When
Construction	Build and verify physical systems	Equipment operates consistently
Fishless Cycle	Establish nitrification	Ammonia/nitrite processing verified
Biological Seeding	Increase biological diversity	Microbial/benthic life expanding
Initial Livestock	Introduce biological demand	Livestock thriving and stable
Nutrient Base	Establish feeding economy	Nutrient trends predictable
Lighting	Bring energy online gradually	Corals respond positively
Operational Verification	Confirm stability	Operating Targets™ consistent

ROS Insight™
The calendar does not commission a reef. Only the ecosystem can demonstrate readiness for the next phase.

Field Note from Allan™
I no longer ask how long to wait. I ask what the reef is telling me. Test kits provide valuable information, but observation provides context.

Common Mistakes

- Assuming the nitrogen cycle ends commissioning.
- Introducing too many organisms too quickly.
- Running full lighting too early.

- Chasing nutrients during early succession.
- Ignoring microbial maturation.
- Relying on the calendar instead of readiness.

Verification

- Nitrogen cycle verified.
- Living Substrate™ biologically active.
- Initial livestock thriving.
- Feeding consistent.
- Lighting introduced deliberately.
- Operating Targets™ stable.

Decision Framework™

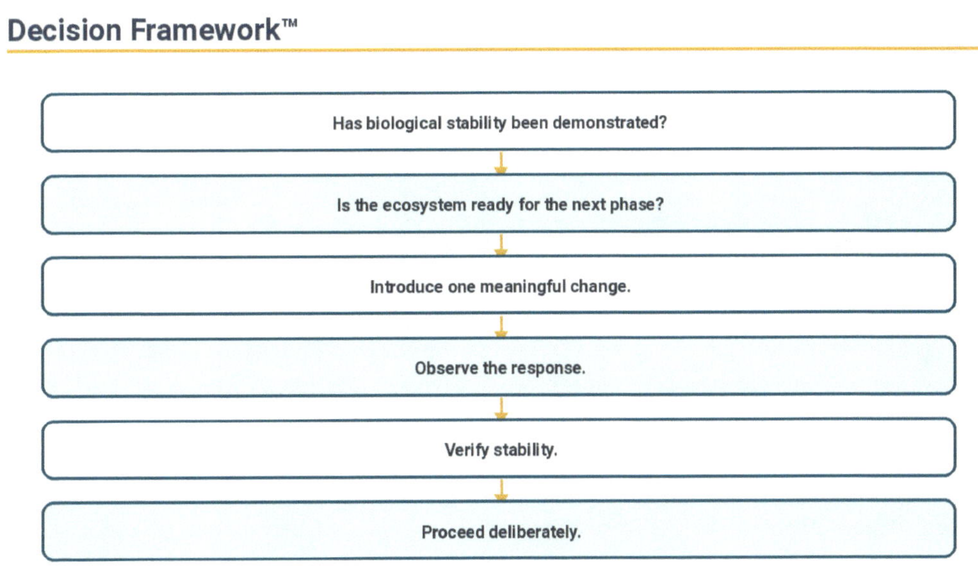

Decision Framework™

Key Takeaways

- Cycling is one milestone within commissioning.
- Commissioning creates a functioning ecosystem.
- Succession continues long after nitrification.
- Lighting, feeding, and livestock are introduced gradually.
- Readiness is biological, not calendar-based.
- Commissioning creates long-term equilibrium.

ROS Principle™

Healthy chemistry is not the goal. It is the signature of a healthy ecosystem.

STANDARD II

Water Chemistry & Equilibrium™

Engineering Chemistry Through Biology Rather Than Chasing Numbers

Purpose

The purpose of this Standard is to establish the principles governing water chemistry within ROS.

Water chemistry is not a collection of independent numbers. It is the chemical expression of biology, gas exchange, lighting, feeding, supplementation, substrate, and flow.

Why This Standard Matters

Measurements are valuable only when interpreted within the ecosystem. Numbers alone cannot explain why conditions exist.

The Ocean Signature™

ROS uses healthy tropical reef conditions as the engineering benchmark. Operating Targets™ include stable salinity, stable temperature, pH generally 8.2–8.4 when practical, natural alkalinity around 7 dKH as a design objective, balanced calcium and magnesium, and appropriate nutrient availability.

Chemistry Reflects Biology

When chemistry changes unexpectedly, ROS first asks what changed in the ecosystem.

pH as Ecosystem Indicator

ROS designs conditions that naturally support pH: gas exchange, indoor air quality, CO_2 management, stable alkalinity, biological activity, and aragonite as part of the larger system.

Alkalinity as an Operational Indicator™

With balanced supplementation, alkalinity often provides early practical insight into biological demand.

Operational Monitoring™

Testing is frequent during commissioning. During long-term operation, testing verifies the Glide Path™. Test strips can support routine monitoring after stability is established. ICP verifies long-term macro and trace trends.

Figure 11-1: The ROS Chemistry Loop™

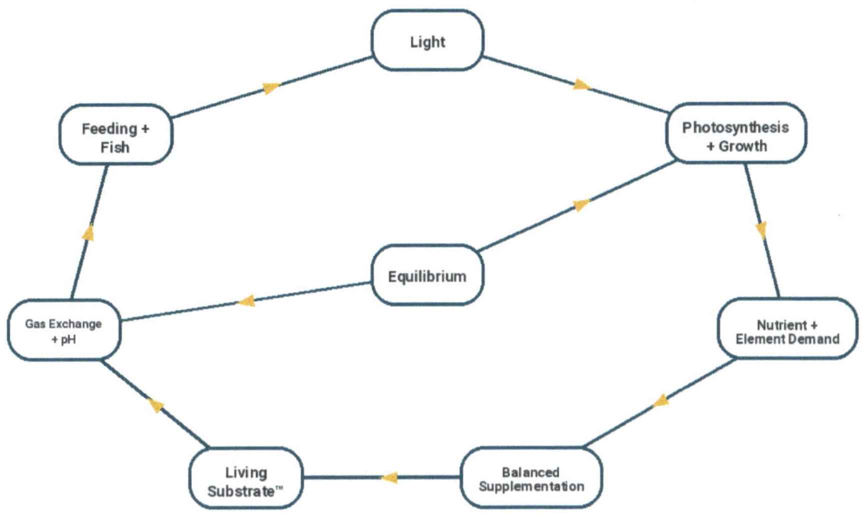

Chemistry is the signature of an ecosystem in operation.

Figure 11-1: The ROS Chemistry Loop™.

ROS Operating Targets™

Parameter	Operating Target™	Operational Note
Temperature	77–79°F / 25–26°C	Stability over exactness
Salinity	35 ppt / ~1.026 SG	Match natural seawater
pH	8.2–8.4 when practical	Engineer through gas exchange and CO2 management
Alkalinity	~7 dKH design objective	Natural reef consistency
Calcium	Natural reef range	Replace consumption
Magnesium	Natural reef range	Maintain balance
Nitrate	Low and stable	Context matters
Phosphate	Low and stable	Trend with nutrient economy

ROS Insight™

The numbers are the instrument panel. They provide information, but the reef provides context.

Field Note from Allan™

One of the biggest changes in my reefkeeping happened when I stopped asking how to raise a number and started asking why the number changed. Often chemistry was not the problem; it was reporting what the biology already

showed.

Common Mistakes

- Chasing isolated parameters.
- Adjusting chemistry without investigating causes.
- Assuming higher numbers are always better.
- Ignoring pH as an ecosystem indicator.
- Supplementing without demonstrated demand.
- Interpreting single measurements without trends.

Verification

- Operating Targets**TM** are stable.
- Coral growth and color remain healthy.
- Supplementation reflects demand.
- Operational Monitoring**TM** confirms consistency.
- ICP supports macro and trace balance.

Decision Framework**TM**

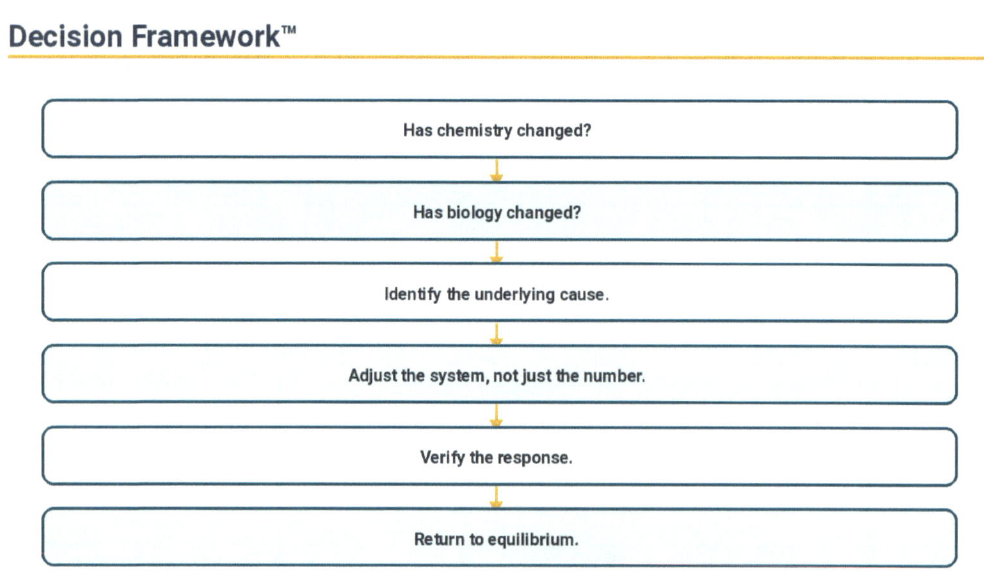

Decision Framework™

Key Takeaways

- Chemistry reflects ecosystem function.
- Operating Targets**TM** support biological success.

- pH and alkalinity are valuable indicators.
- Replace what growth consumes.
- Operational Monitoring™ and ICP complement one another.
- Healthy chemistry is engineered through design and stewardship.

ROS Principle™

Engineer the ecosystem, and chemistry becomes its signature.™

Standard 12: Automation & Consistency

Automation succeeds when the reef notices consistency—not technology.

STANDARD 12

Automation & Consistency™

Using Technology to Improve Repeatability, Not Complexity

Purpose

The purpose of this Standard is to establish the role of automation within ROS.

Automation is implemented when it improves consistency, reduces human variability, and supports long-term stability.

Why This Standard Matters

Natural reefs are predictable. Aquariums often become variable because humans are variable. Thoughtful automation reduces unnecessary variation.

The ROS Automation Philosophy

Automation should answer: does this improve consistency while supporting equilibrium?

Prioritize Repeatability

Automate tasks where consistency directly affects stability: top-off, lighting, feeding, dosing, temperature control.

Micro-Dosing

Frequent small-volume supplementation better matches continuous biological consumption and reduces short-term fluctuation.

Redundancy

Critical systems deserve risk management: temperature, top-off, circulation, battery backup, and leak detection where appropriate.

Automation Does Not Replace Stewardship

Technology performs tasks. The aquarist remains responsible for observation, judgment, and ethical care.

Figure 12-1: The ROS Automation Pyramid™

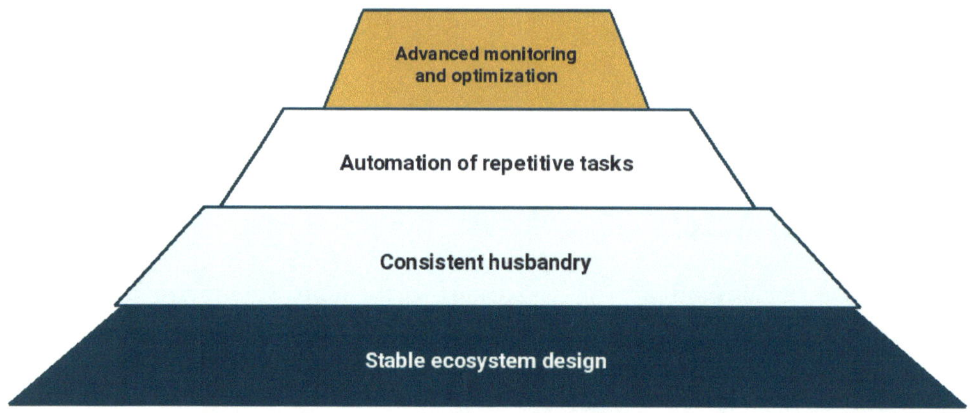

Automation builds upon good design. It cannot replace it.

Figure 12-1: The ROS Automation Pyramid™.

Figure 12-2: Automated Tasks vs. Human Stewardship

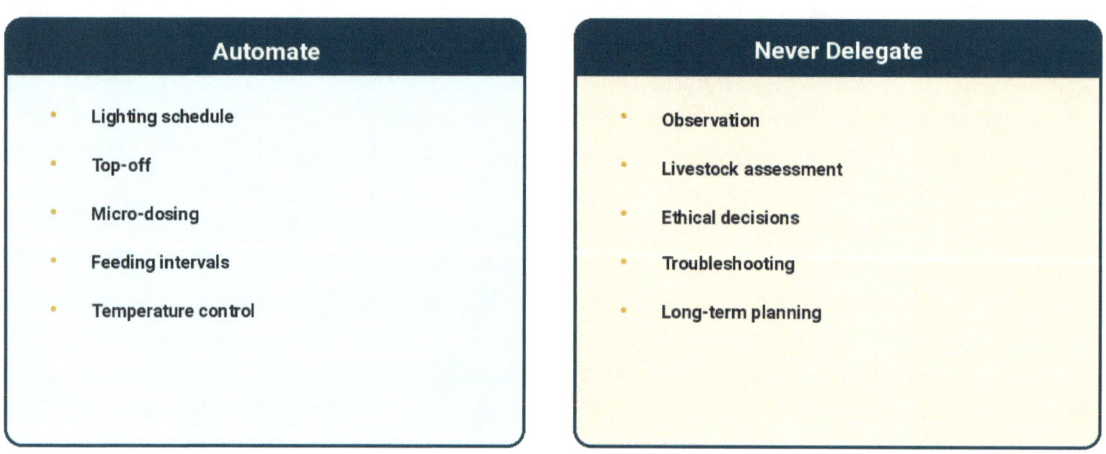

Figure 12-2: Automated Tasks vs. Human Stewardship.

Automation Priority Table

Task	ROS Value	Failure Planning

Automatic top-off	Stable salinity	Overfill prevention, reservoir check
Lighting schedule	Stable photoperiod	Timer/controller backup
Automatic feeding	Consistent nutrient input	Avoid overfeeding if jammed
Micro-dosing	Stable replacement of consumption	Limit reservoir volume, calibrate
Temperature control	Thermal stability	Redundant thermostat, alarms
Monitoring	Awareness	Do not replace observation

ROS Insight™

The best automation is almost invisible. It quietly performs routine tasks while the reef remains the center of attention.

Field Note from Allan™

In aviation, automation reduces workload only when the pilot understands it. Reef automation should work the same way: it provides consistency while stewardship remains human.

Common Mistakes

- Automating tasks that do not benefit from automation.
- Assuming automation replaces observation.
- Overcomplicating simple systems.
- Failing to maintain automated equipment.
- Depending on internet for critical life support.
- Ignoring failure planning.

Verification

- Routine tasks occur consistently.
- Human error is reduced.
- Equipment remains simple to maintain.
- Critical systems have redundancy.
- The aquarist observes more than operates equipment.

Decision Framework™

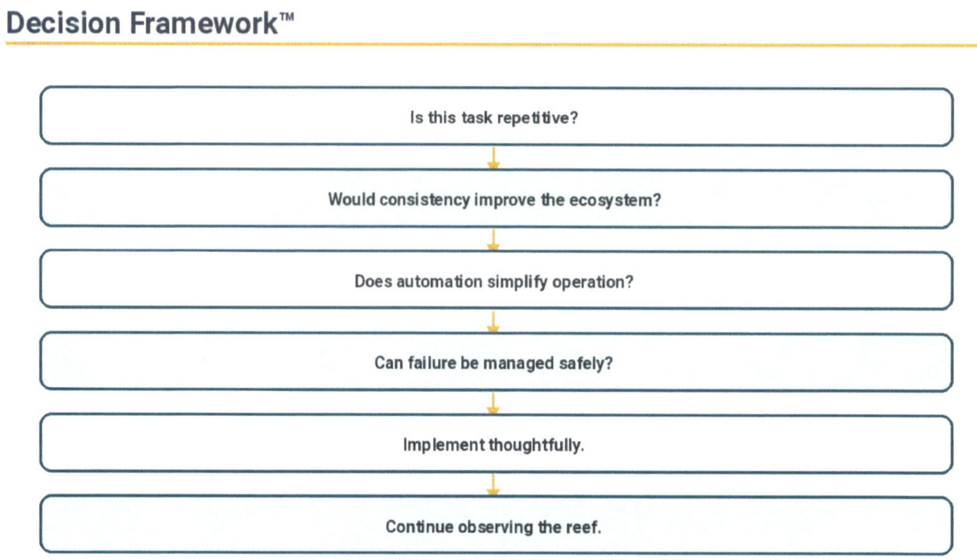

Decision Framework™

Key Takeaways

- Automation improves consistency, not prestige.
- Repetitive tasks benefit most.
- Micro-dosing aligns with continuous demand.
- Redundancy protects critical systems.
- Automation never replaces stewardship.
- The simplest effective solution is preferred.

ROS Principle™
Automate consistency. Never automate stewardship.™

Standard 13: Troubleshooting Through ROS

Every symptom is information. Very few symptoms are the problem.

STANDARD 13

Troubleshooting Through the ROS Framework™

Diagnosing Systems Rather Than Treating Symptoms

Purpose

The purpose of this Standard is to establish the diagnostic methodology of ROS.

Troubleshooting is the disciplined process of identifying the underlying cause of change within a living ecosystem.

Why This Standard Matters

Every reef changes. Successful reefkeepers interpret these changes before reacting. ROS teaches diagnostic thinking.

The ROS Diagnostic Philosophy

Observation, measurement, understanding, action, verification. The order matters.

The Equilibrium Test™

Before making a significant adjustment, ask whether the decision will move the ecosystem closer to Operating Targets™ or simply react to a symptom.

Diagnose the System

Review lighting, flow, gas exchange, feeding, livestock, substrate, supplementation, temperature, salinity, and monitoring history.

Root Cause Before Remedy

Low pH may involve CO_2 and gas exchange. Algae may involve nutrients, lighting, maturity, herbivores, and flow. Coral color changes may involve PAR, flow, nutrition, chemistry, or disease.

Figure 13-1: The ROS Diagnostic Tree™

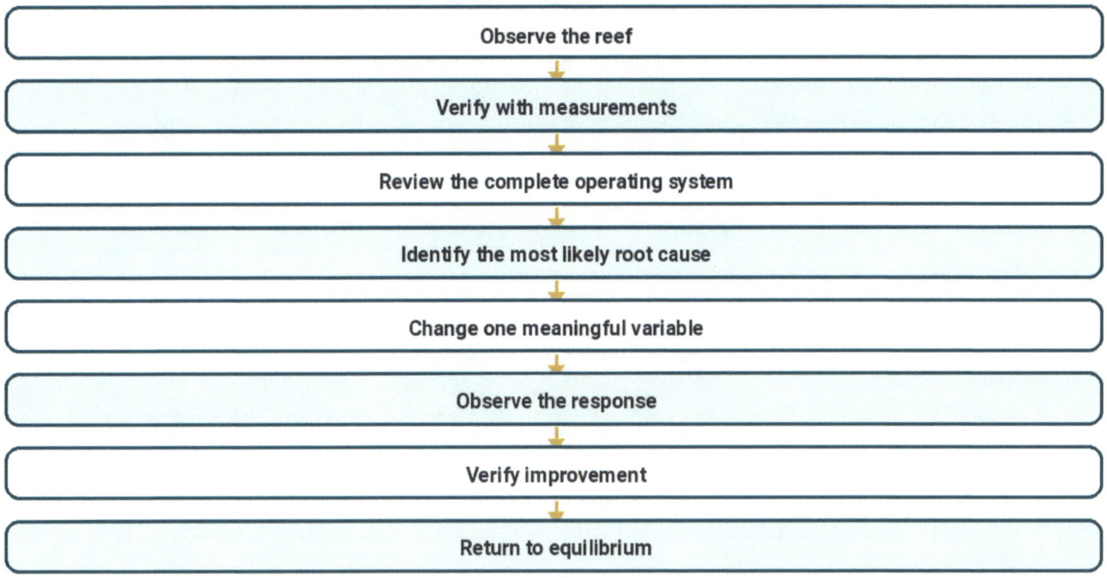

Figure 13-1: The ROS Diagnostic Tree™.

ROS Insight™
Every symptom is part of a story. The story emerges when fish, corals, chemistry, water clarity, flow, and history are considered together.

Field Note from Allan™
Pilots diagnose before they react. Physicians treat patients, not isolated lab values. The reef deserves the same respect: understand what it is telling you before trying to solve it.

Common Mistakes

- Reacting to isolated test results.
- Changing multiple variables simultaneously.
- Treating symptoms instead of causes.
- Ignoring trends.
- Assuming every change requires immediate intervention.
- Forgetting biology responds slowly.

Verification

- Troubleshooting follows a consistent process.
- Root causes are identified before corrections.
- Interventions are less frequent and more effective.

- Records reveal trends.
- The reef returns to equilibrium with minimal disruption.

Decision Framework™

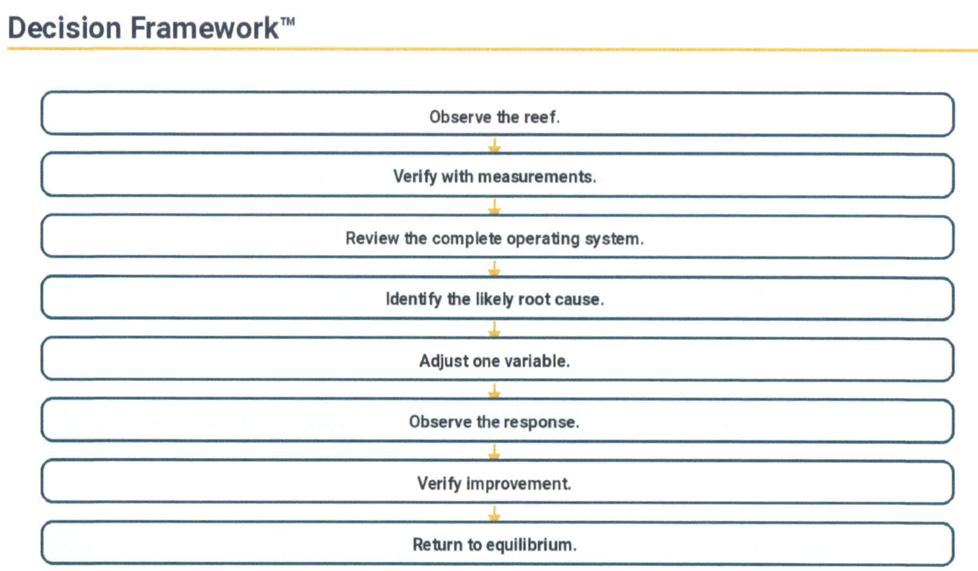

Decision Framework™

Key Takeaways

- Troubleshooting begins with observation.
- Symptoms require context.
- Root causes deserve more attention than symptoms.
- Trends are more valuable than isolated values.
- One deliberate change teaches more than many simultaneous changes.
- Healthy reefs need fewer corrections.

> **ROS Principle™**
> **Diagnose the ecosystem. Correct the cause. Verify the outcome.™**

A mature reef does not demand attention. It rewards observation.

STANDARD 14

Achieving Equilibrium™

Recognizing When the Reef Begins Operating Naturally

Purpose

The purpose of this Standard is to define the characteristics of a mature Reef Operating System™ operating in biological equilibrium.

Equilibrium does not mean nothing changes. It means natural changes occur predictably, gradually, and without continual correction.

Why This Standard Matters

Many reefkeepers pursue stability without defining it. ROS defines success as predictable, healthy operation that feels resilient rather than fragile.

Characteristics of Equilibrium

Biological stability, chemical stability, physical stability, and operational stability are interpreted together.

The Reef Begins Self-Stabilizing™

Mature ecosystems increasingly compensate for routine variation. This is not maintenance-free; it is naturally stable.

The Glide Path™

Equilibrium is a continuous glide path. Adjustments become smaller, less frequent, and increasingly predictable.

Growth Becomes the Primary Consumer

As equilibrium develops, healthy growth drives consumption. The operating objective becomes replacing what growth consumes.

Figure 14-1: Signs of a Self-Stabilizing Reef™

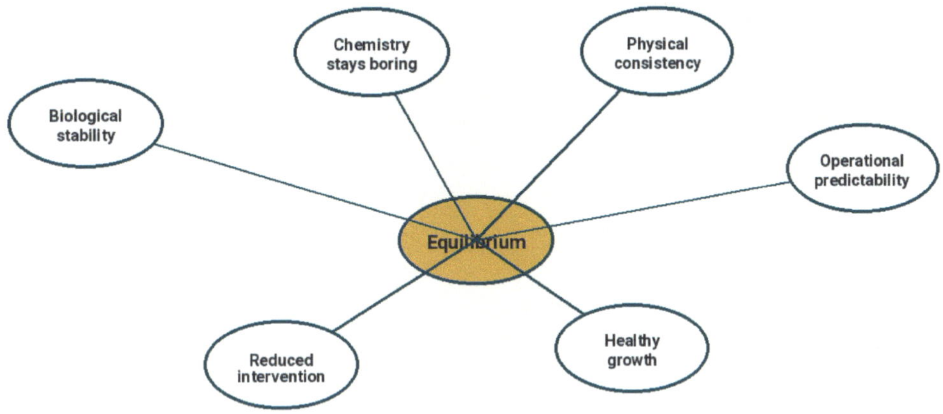

Figure 14-1: Signs of a Self-Stabilizing Reef™.

Equilibrium Indicators

System Area	Signs of Equilibrium
Biology	Natural behavior, steady coral extension, healthy coloration
Chemistry	Stable Operating Targets™, predictable alkalinity consumption
Physical	Stable temperature, effective flow, active Living Substrate™
Operation	Routine stewardship, less troubleshooting, fewer large interventions

ROS Insight™

Stability creates freedom: freedom from constant troubleshooting and freedom to enjoy the reef as a living ecosystem.

Field Note from Allan™

One day you realize weeks have passed without needing to fix anything. The fish greet you, corals grow, chemistry holds steady, and maintenance is predictable. To me, that is equilibrium.

Common Mistakes

- Believing equilibrium means perfection.
- Confusing temporary stability with maturity.
- Optimizing an already stable system.
- Introducing unnecessary changes.
- Mistaking routine stewardship for instability.

Verification

- The reef consistently maintains Operating Targets**TM**.
- Coral growth remains steady and natural.
- Biological demand changes gradually.
- Routine stewardship replaces troubleshooting.
- The reef demonstrates resilience.

Decision Framework**TM**

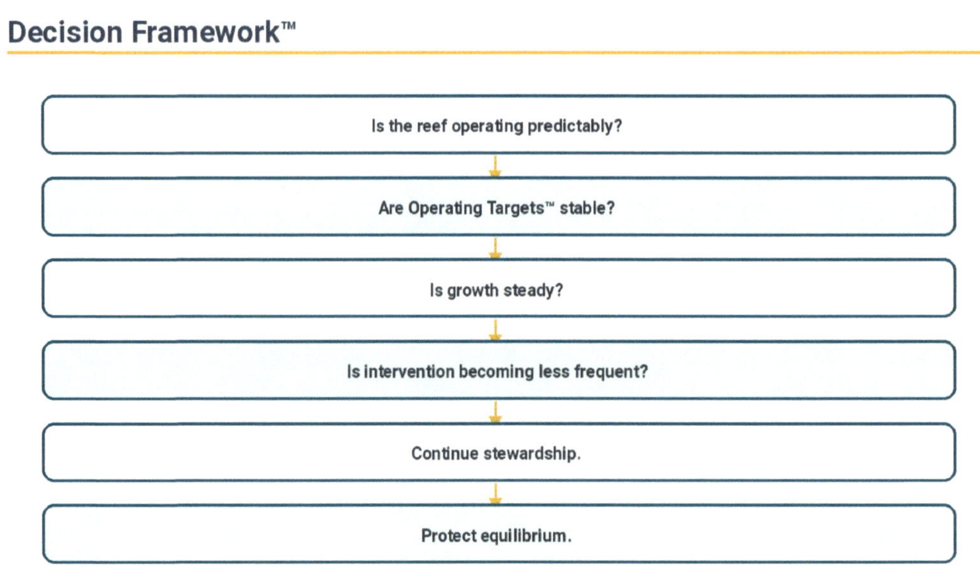

Decision Framework™

Key Takeaways

- Equilibrium is dynamic, not static.
- A Self-Stabilizing Reef**TM** remains biologically balanced.
- Growth becomes the primary consumer of supplementation.
- Mature reefs require stewardship, not repair.
- Resilience indicates long-term success.
- The goal is predictable healthy operation.

ROS Principle™
Protect equilibrium, and the reef will reward your patience.™

The Defining ROS Statement

The goal of the Reef Operating System™ is not to force an aquarium into acceptable parameters.

The goal is to design an ecosystem that naturally maintains conditions characteristic of healthy tropical coral reefs, requiring only the replacement of materials that are permanently consumed by growth.

A mature reef is never finished. It simply becomes better.

STANDARD 15

Continuous Improvement™

Refining the Ecosystem Without Disrupting Equilibrium

Purpose

The purpose of this Standard is to establish the principles governing continuous improvement within ROS.

Maturity is not the end of development. It begins a phase of deliberate refinement that protects equilibrium.

Why This Standard Matters

Healthy ecosystems evolve. Technology improves. Aquarists learn. ROS encourages improvement without sacrificing the stability that made improvement possible.

Improvement Begins with Observation

Observe coral growth, fish behavior, nutrient trends, supplementation, equipment performance, maintenance efficiency, and resilience.

Improve One System at a Time

Define the objective, change one variable, observe, verify, document, and learn.

Technology Evolves

ROS separates principles from products. Better tools may be adopted when they improve biology, consistency, simplicity, or equilibrium.

Data Becomes Knowledge

Years of operational history reveal seasonal variation, growth trends, consumption, equipment longevity, and livestock development.

Figure 15-1: Continuous Improvement Loop™

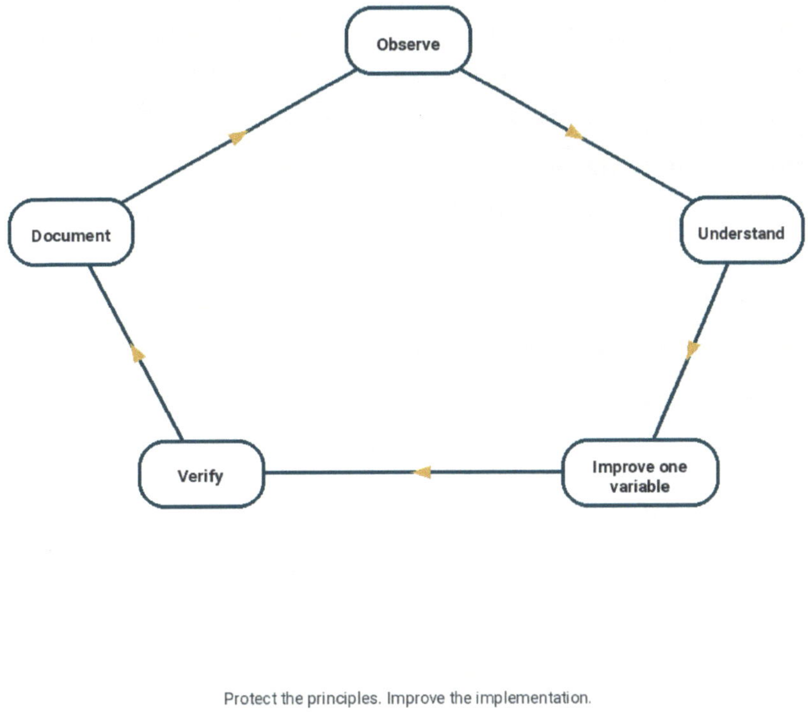

Figure 15-1: The Continuous Improvement Loop™.

Improvement Log Template

Date	Objective	Change	Verification	Result

ROS Insight™

Stability makes innovation possible. Predictable systems make the effect of thoughtful improvements easier to see.

Field Note from Allan™

Every flight teaches something. Every patient teaches something. Every reef teaches something. The goal is to make tomorrow's system better while protecting the life entrusted to our care.

Common Mistakes

- Changing because something new is available.
- Abandoning proven practices without benefit.

- Changing multiple systems at once.
- Failing to document improvements.
- Mistaking novelty for progress.
- Forgetting stability is an achievement.

Verification

- Improvements are deliberate and measurable.
- Records support decisions.
- Equilibrium is preserved.
- New technologies are evaluated objectively.
- The reef becomes more resilient rather than more complicated.

Decision FrameworkTM

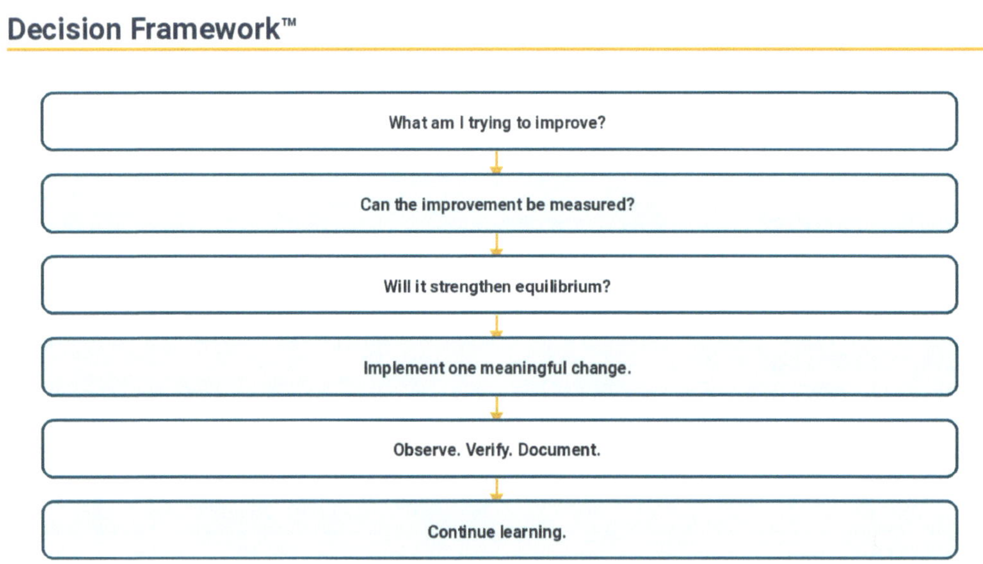

Decision Framework™

Key Takeaways

- Continuous improvement begins with observation.
- Stability enables meaningful innovation.
- Principles remain constant as products evolve.
- One measured improvement is better than many unverified changes.
- Records transform experience into knowledge.
- Stewardship includes responsible sharing.

ROS Principle™

Protect the principles. Improve the implementation.™

Standard 16: The Steward's Commitment

We do not own the reef. We are entrusted with its care.

STANDARD 16

The Steward's Commitment™

The Responsibility of Caring for Living Ecosystems

Purpose

The purpose of this Standard is to define the responsibilities that accompany the practice of ROS.

Knowledge alone does not create exceptional reefkeepers. Stewardship does.

Why This Standard Matters

A reef aquarium is not furniture or decoration. It is a living ecosystem composed of organisms that depend entirely on the environment created by the aquarist.

The Stewardship Philosophy

A successful steward designs thoughtfully, observes carefully, intervenes deliberately, learns continuously, acts responsibly, shares knowledge, and respects nature.

Stewardship Beyond the Aquarium

Support ethical collection and aquaculture, responsible husbandry, new hobbyists, honest observation, evidence, and respectful discussion.

Humility Before Nature

No aquarium perfectly reproduces the ocean. That is not failure. It is inspiration. Nature remains the benchmark.

Leave the Reef Better

Every decision should ask whether it leaves the ecosystem healthier than it was found.

Stewardship Commitments

Commitment	Practical Expression
Animal welfare	Select livestock responsibly and plan for adulthood
Humility	Let nature remain the benchmark
Discipline	Observe before intervening
Evidence	Measure before concluding
Community	Share knowledge honestly

Continuity	Continue learning and improving

ROS Insight™
The ocean is still the standard. Every aquarium is an interpretation of it, and every observation should make us better stewards.

Field Note from Allan™
Flying taught discipline. Medicine taught compassion. Business taught leadership. Travel taught perspective. Raising five children taught patience. Reefkeeping brought these lessons together: healthy systems are guided through understanding, not force.

Common Mistakes

- Prioritizing appearance over animal welfare.
- Pursuing novelty instead of long-term health.
- Confusing ownership with stewardship.
- Assuming experience ends learning.
- Forgetting each decision affects living organisms.

Verification

- Animal welfare guides decisions.
- The ecosystem becomes healthier over time.
- Knowledge is shared responsibly.
- Continuous learning becomes habit.
- Stewardship defines the aquarist.

Decision Framework™

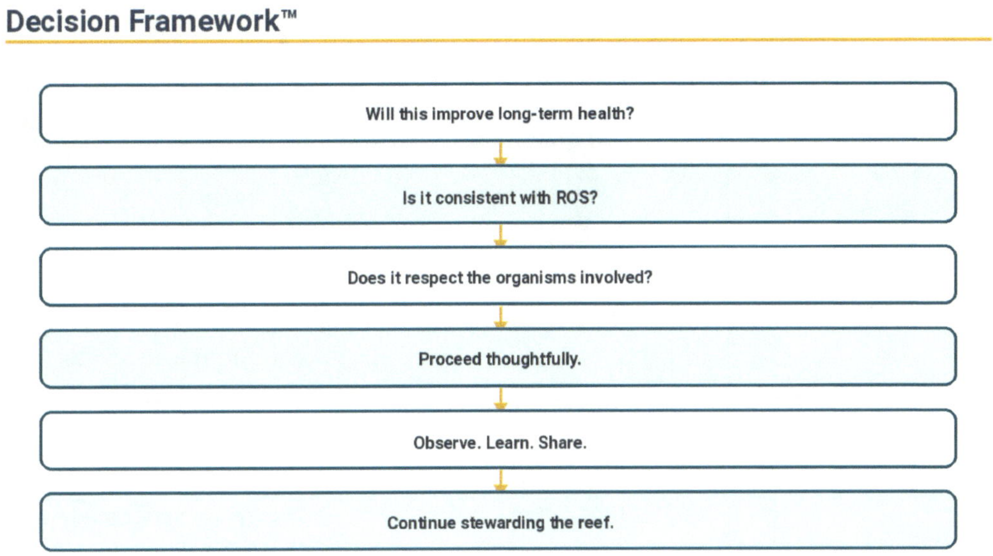

Decision Framework™

Key Takeaways

- Stewardship defines ROS.
- Every aquarium is a responsibility.
- Nature remains the benchmark.
- Learning is an obligation.
- Healthy reefs inspire communities.
- The legacy of ROS is measured by the stewards it creates.

ROS Principle™
Steward the reef with humility, discipline, and respect for the life entrusted to your care.™

Appendix A: Operating TargetsTM

Ocean SignatureTM Reference Table

Area	Operating Target™	Operational Interpretation
Temperature	77–79°F / 25–26°C	Stability is more important than exact value.
Salinity	35 ppt / approx. 1.026 SG	Maintain with ATO and verification.
pH	8.2–8.4 when practical	Engineer through gas exchange, CO2 control, and stability.
Alkalinity	Approx. 7 dKH design objective	Natural reef consistency; monitor as operational indicator.
Calcium	Natural reef range	Verify periodically; replace consumption.
Magnesium	Natural reef range	Maintain balance with supplementation.
Nitrate	Low, stable, context-dependent	Nutrients are resources; trends matter.
Phosphate	Low, stable, context-dependent	Interpreted with nitrate, feeding, and biology.

Biological Operating TargetsTM

Component	Desired Operating Condition
Fish	Natural behavior, healthy body condition, normal respiration, consistent feeding.
Corals	Steady growth, natural coloration, normal polyp extension, colony development.
Living Substrate™	Active clean-up crew, no chronic stagnation, biologically healthy appearance.
Ecosystem	Predictable, stable, resilient, and increasingly self-stabilizing.

Appendix B: Equipment & Build Specifications™

ROS Equipment Decision Matrix

Component	ROS Status	Typical Trigger	Notes
Aquarium Platform	Required	Always	Choose for intended ecosystem.
Lid	Preferred	Evaporation, temperature, fish safety	Passive stability.
Modified UGF / Plenum	Preferred core implementation	Living Substrate™ support	Uniform distributed flow.
CO2 Scrubber	Preferred when CO2 limits pH	Low pH / indoor CO2	Supports natural gas chemistry.
Protein Skimmer	Compatible option	High bioload, gas exchange, export	Tool, not requirement.
ATO	Preferred	Salinity variation	Consistency.
Doser	Preferred	Supplement replacement	Hourly micro-dosing when practical.
Controller	Compatible option	Monitoring, safeguards	Never replaces stewardship.
UV	Compatible option	Water clarity / disease strategy	Use with objective.
Sump	Compatible option	Water volume / equipment organization	Not required.

Living Substrate Build Reference

Item	Specification	Purpose
Substrate	Aragonite-based media	Chemical compatibility and microbial surface.
Depth	1/2 to 1 1/2 inches preferred	Biological function with manageable stewardship.
Screen	Patio screen for fine substrates	Protects UGF plate.
Flow	Approx. 50 GPH distributed across bottom	Design objective; uniformity matters most.
Cleanup Crew	Ceriths and function-matched species	Surface maintenance and biodiversity.

Appendix C: Commissioning Checklist™

ROS Commissioning Checklist

Step	Action	Verified
1	Platform level, stable, and accessible	☐
2	Modified UGF plate installed	☐
3	Patio screen installed if needed	☐
4	Aragonite substrate added	☐
5	Uniform under-substrate flow verified	☐
6	Temperature stable	☐
7	Salinity verified	☐
8	Gas exchange / CO2 scrubber configured	☐
9	Fishless biological cycle started	☐
10	Ammonia/nitrite processing verified	☐
11	Biological seeding introduced	☐
12	Initial livestock quarantined/cleared	☐
13	Nutrient base established	☐
14	PAR measured and lighting commissioned	☐
15	Operational Monitoring™ baseline recorded	☐

Appendix D: Operational Monitoring™ Sheets

Weekly Operational Monitoring™

Date	Temp	Salinity	pH	Alk	Nitrate Trend	Observation Notes

Monthly Equilibrium Review

Question	Yes/No	Notes
Is pH on Glide Path™?		
Is alkalinity stable?		
Is coral growth natural?		
Is feeding consistent?		
Has anything changed?		
Is intervention needed?		

Appendix E: Diagnostic Decision Trees™

Diagnostic Tree: Low pH

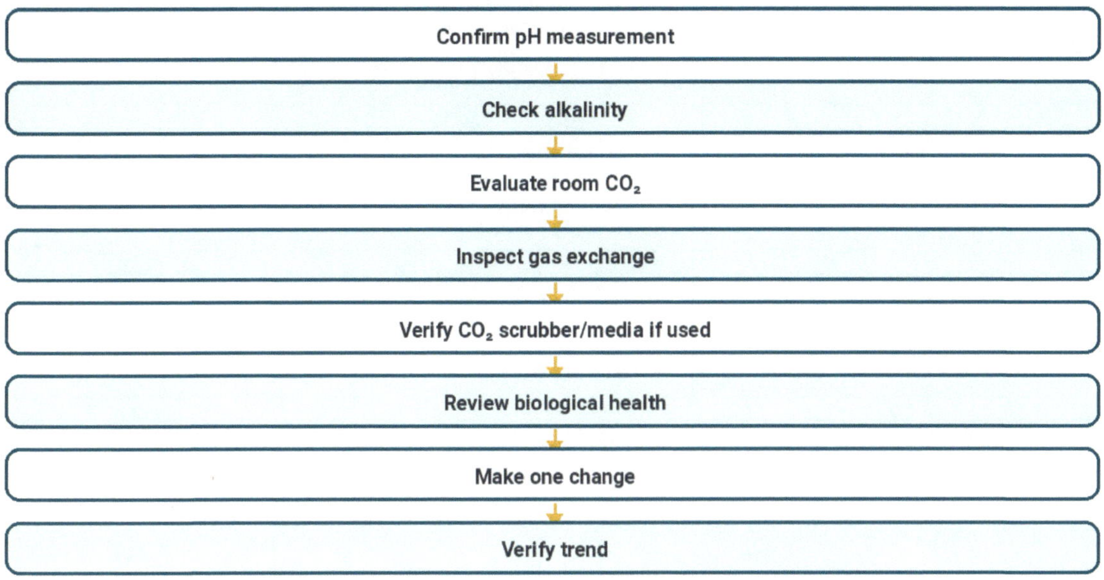

Diagnostic Tree: Low pH

Diagnostic Tree: Algae Growth

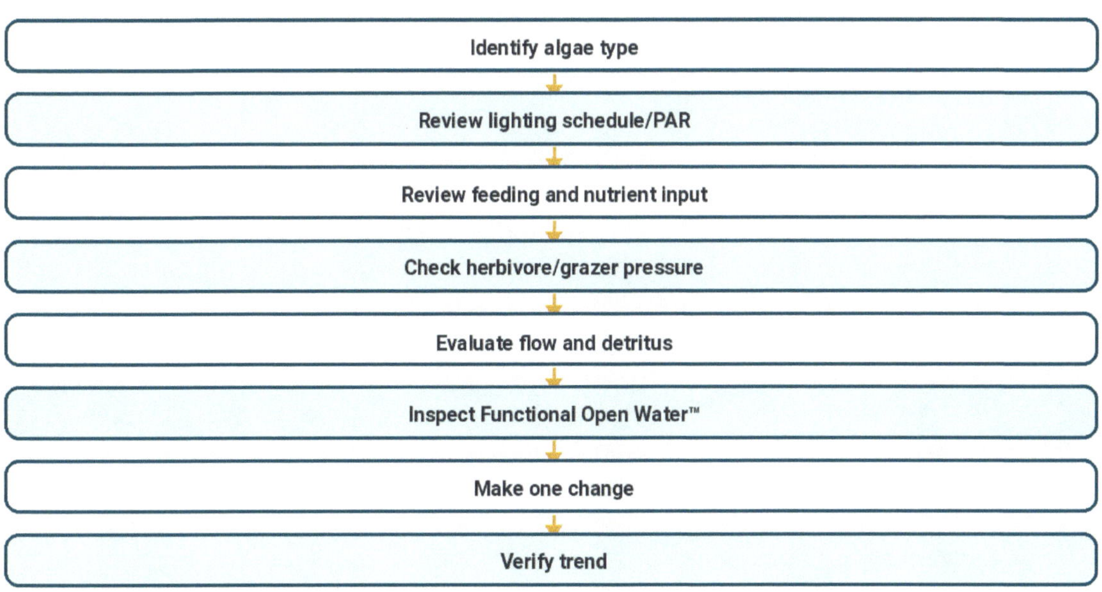

Diagnostic Tree: Algae Growth

Diagnostic Tree: Poor Coral Color

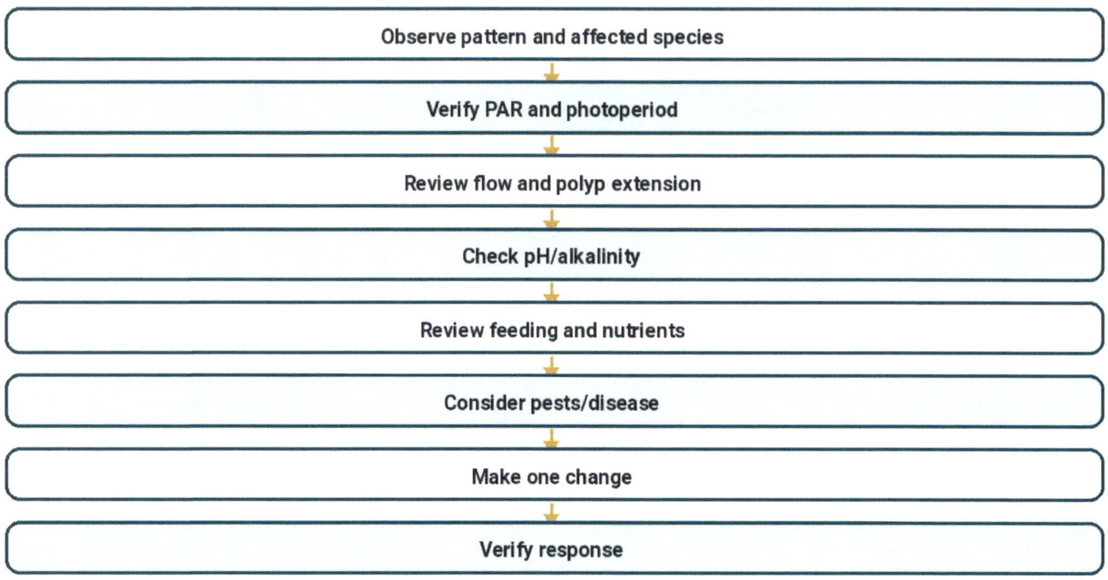

Diagnostic Tree: Poor Coral Color

Appendix F: ROS Lexicon TM

ROS Lexicon TM

Term	Definition
Reef Operating System™	The complete ecosystem design methodology.
Equilibrium Method™	The core philosophy of designing aquariums that naturally seek stability.
Ocean Signature™	The defining operating characteristics of healthy tropical reefs used as a benchmark.
Operating Targets™	Design objectives used to evaluate ecosystem performance.
Glide Path™	The predictable trajectory of a system operating toward equilibrium.
Living Substrate™	The engineered biological substrate system used as a foundation of ROS.
Functional Open Water™	Open water intentionally preserved for flow, fish behavior, light, and future growth.
Ecosystem Commissioning™	The process of bringing a reef ecosystem into biological operation.
Operational Monitoring™	Simplified routine verification after stability is established.
Self-Stabilizing Reef™	A mature reef that remains stable through biological equilibrium and stewardship.
Feed the Ecosystem™	The principle that food nourishes the entire system, not only individual animals.
Minimum Necessary Complexity™	The discipline of adding complexity only when it produces measurable benefit.
Biology Before Technology™	The principle that technology should support biological processes, not replace them.
Stewardship	The ongoing responsibility to care for the living ecosystem.

Figure List

Visual Credits & Illustration Statement

The current pre-publication proof uses AI-assisted illustrative visual plates, aquarium scenes, cover artwork, and original ROS diagrams for editorial design purposes. These visuals are intended to communicate aspiration, beauty, structure, and ecological imagination. They are not presented as species-identification photography, site-specific reef documentation, or scientific field records. Any future owner-provided, licensed, or public-domain photography should be documented in the internal rights log before final release.

References & Scientific Support

The Reef Operating System**TM** is an applied reef-aquarium methodology developed by Allan Poulin. It integrates principles from coral reef ecology, marine chemistry, aquarium husbandry, systems thinking, and long-term observation.

The references below support the scientific background underlying the Specification. ROS terminology, Standards, decision frameworks, diagrams, checklists, operating methodology, and educational tools remain original Reef Operating System**TM** materials.

Coral Reef Ecology & Biodiversity

These sources support the ROS emphasis on treating the aquarium as a living ecosystem rather than as a collection of unrelated organisms or equipment.

- National Oceanic and Atmospheric Administration. "Coral Reef Ecosystems." NOAA Education.
- United States Environmental Protection Agency. "Basic Information about Coral Reefs."
- NOAA Atlantic Oceanographic and Meteorological Laboratory. "Coral Reef Ecosystems."
- United Nations Environment Programme. "Coral Reefs."

Seawater Chemistry, Carbon Dioxide & pH

These sources support the ROS emphasis on gas exchange, carbon dioxide management, pH stability, alkalinity awareness, and carbonate chemistry as components of reef-aquarium equilibrium.

- National Oceanic and Atmospheric Administration. "Ocean Acidification."
- NOAA Pacific Marine Environmental Laboratory. "What Is Ocean Acidification?"
- NOAA Ocean Acidification Program. Ocean Acidification Data and Monitoring Resources.

Reef Aquarium Chemistry & Husbandry

These references support the ROS sections on Ocean Signature**TM**, Operating Targets**TM**, balanced supplementation, alkalinity as an operational indicator, and replacing what growth permanently consumes.

- Holmes-Farley, Randy. "Reef Aquarium Water Parameters." Reefkeeping Magazine.
- Holmes-Farley, Randy. "Calcium and Alkalinity." Reefkeeping Magazine.
- Holmes-Farley, Randy. "The Relationship Between Alkalinity and pH." Reefs Magazine.
- Holmes-Farley, Randy. "Magnesium in Reef Aquaria." Reefs Magazine.
- Holmes-Farley, Randy. "How a Two-Part Alkalinity and Calcium System Works, and Why It Matters." Reef2Reef.
- Reefkeeping Magazine. "Chemistry Archive."

Additional Book References to Consider

- Delbeek, J. Charles, and Julian Sprung. The Reef Aquarium, Volumes 1-3.
- Borneman, Eric. Aquarium Corals.
- Adey, Walter H., and Karen Loveland. Dynamic Aquaria.
- Fenner, Robert. The Conscientious Marine Aquarist.
- Rohwer, Forest, and Merry Youle. Coral Reefs in the Microbial Seas.

Exact edition details should be verified before final publication layout.

Visual Asset Classification

The visual assets in this Specification are classified for publication purposes as follows. A more detailed internal rights log should be maintained as part of the production record.

Asset Type	Use in Specification	Rights Status
Cover artwork	Editorial cover design and visual positioning	Original AI-assisted illustrative artwork
Section-opening reef images	Editorial visual plates	Original AI-assisted illustrative artwork
Aquarium display images	Editorial visual plates and concept scenes	Original AI-assisted illustrative artwork
Diagrams and diagnostic trees	Instructional educational figures	Original ROS educational materials
Tables and checklists	Operating references and implementation tools	Original ROS educational materials
Icons and process graphics	Visual navigation and instructional support	Original ROS educational materials
Future owner-provided photography	Optional replacement or supplemental imagery	To be documented by source
Future licensed photography	Optional replacement or supplemental imagery	To be documented by license

Internal Rights Log Fields

Field	Purpose
Asset ID	Unique number for each image or figure
Page number	Where the asset appears
Asset description	What the asset depicts
Asset type	AI-assisted, diagram, table, owner photo, licensed photo, or public domain
Source	Where the asset came from
License / rights status	Why it is usable
Attribution required	Yes or no
Notes	Limitations, replacement plans, or production comments

Final Pre-Publication Checklist

This checklist governs the final go / no-go decision before Amazon KDP upload and direct PDF release.

Manuscript Completion

- All 16 Standards are included.
- Front matter, appendices, ROS Lexicon, Figure List, references, notices, and disclaimer are complete.
- Edition is consistently labeled Revision A.

Design & Layout

- Deep coastal color palette is applied throughout.
- Diagram text fits inside boxes.
- Diagnostic trees, tables, and dashboards are readable.
- Cover design matches the interior theme.

Image & Rights Review

- AI-assisted imagery is disclosed as illustrative.
- No generated image is described as a real location or factual scientific photograph.
- The internal visual asset log is maintained.

Scientific & Legal Support

- References section is included.
- Strong claims are defensible and appropriately qualified.
- No guarantee of livestock survival or financial outcome is implied.
- Publication Disclaimer and Responsible Husbandry Notice are included.

Cover Production

- Paperback and hardcover covers are produced separately.
- ISBN/barcode placeholders are present.
- Spine width matches final page count and production specs.
- Bleed and safe margins are correct.

ISBN & Imprint

- Paperback ISBN assigned.
- Hardcover ISBN assigned.
- Optional digital ISBN assigned.
- Publisher / imprint listed as Reef Operating System Press.

Amazon KDP Metadata

- Title, subtitle, edition, author, keywords, categories, and pricing are locked.
- Product description matches premium reference positioning.

Proofing

- Paperback proof copy ordered and inspected.
- Hardcover proof copy ordered and inspected.
- PDF proof reviewed on desktop and tablet.
- Margins, cover trim, spine alignment, diagram readability, and color consistency are verified.

Direct Digital Sales

- Premium PDF file prepared.
- Sales page, checkout, delivery, refund policy, and licensing terms prepared.

Publication Metadata

Field	Locked Value
Title	The Reef Operating System™ Specification
Subtitle	Engineering Natural Balance for Thriving Marine Reef Aquariums
Edition	Revision A
Author	Allan Poulin
Author Descriptor	Aviator · Systems Thinker · Husband · Father · Reef Steward
Imprint	Reef Operating System Press
Formats	Premium Digital PDF, Paperback, Hardcover
Launch Keywords	no water change reef tank; low maintenance reef aquarium; reef tank setup for beginners; saltwater aquarium for beginners; how to start a reef tank; coral reef aquarium guide; reef tank water chemistry
Premium PDF Founder's Edition	299 USD
Paperback Revision A	129.95 USD
Hardcover Revision A	199.95 USD
Hardcover ISBN	979-8-9969713-0-5
Paperback ISBN	979-8-9969713-1-2

www.ingramcontent.com/pod-product-compliance
Lightning Source LLC
LaVergne TN
LVHW070936210926
848482LV00003B/259
* 9 7 9 8 9 9 6 9 7 1 3 1 2 *